Chipless RFID Authentication

Series Editor
Etienne Perret

Chipless RFID Authentication

Design, Realization and Characterization

Zeshan Ali
Etienne Perret
Nicolas Barbot
Romain Siragusa

iSTE

WILEY

First published 2022 in Great Britain and the United States by ISTE Ltd and John Wiley & Sons, Inc.

ISTE Ltd
27-37 St George's Road
London SW19 4EU
UK

www.iste.co.uk

John Wiley & Sons, Inc.
111 River Street
Hoboken, NJ 07030
USA

www.wiley.com

Library of Congress Control Number: 2022936226

British Library Cataloguing-in-Publication Data
A CIP record for this book is available from the British Library
ISBN 978-1-78630-833-7

Contents

Preface

Counterfeiting has become a global and dynamic phenomenon, as in 2013 the total international trade of counterfeited items was up to 2.5% of the global trade. This illicit practice poses threat to a wide range of industries and harms societies from various perspectives: ultraexpensive consumer goods (e.g. cosmetics, fragrances, leather articles, jewelry), business-to-business goods (e.g. tools, appliances, materials, replacement parts) and essential consumer goods (e.g. food items, medicines). Product authentication offers vast opportunities to combat fakes in the global supply chain. Therefore, robust and reliable authentication methods have become a global demand to limit counterfeiting.

This book is focused on taking the next step with the aim of developing chipless tags for highly secure product authentication applications. The concept of conventional chipless radio frequency identification (RFID) is extended to the authentication where each tag has to present a unique signature that can never be reproduced even if someone tries to copy the tag. For this purpose, natural randomness (i.e. inherent in the fabrication process) along the dimensional parameters of resonators is used. Such natural randomness can produce unique electromagnetic (EM) signatures that can be used for authentication. First, a methodology to characterize the chipless RFID tag for authentication applications is presented. This methodology consists of procedures to conduct both authentication and statistical analyses. The capabilities of chipless technology to be used for tag discrimination are demonstrated by purposely applying the dimensional variations using two technologies: printed circuit board (PCB) and inkjet printing. Then, the extraction of aspect-independent parameters for chipless RFID tags is presented. For authentication purposes, aspect-independent parameters are directly associated with the physical dimensions of the scatterer of a chipless tag, but not associated with the measurement procedure. The random variation of the physical dimensions of the scatterers is then associated with aspect-independent parameters, which is particularly promising for chipless authentication. On the other hand, with the operation of a single measurement, the proposed extraction of aspect-independent parameters is very promising for the practical implementation of the chipless RFID technology. Finally, chipless authentication methods using naturally occurring randomness in the realization process of PCB chipless tags and inkjet-printed polyethylene terephthalate (PET) chipless tags are presented. The optimization of chipless RFID tags for each realization technology (PCB and inkjet printing) is presented. This optimization is performed to exploit the natural process variations effectively for the purpose of authentication, unlike the conventional chipless RFID tags that are not capable of exploiting the variations effectively. To prove this concept, sufficiently large populations of chipless RFID tags are taken. For PCB, chipless RFID tags are realized two times intermittently, where each realization consists of 45 tags. The two different realizations share the same company, the same PCB technology, but a different film mask, in order to ensure the natural dimensional randomness. Similarity analyses are conducted within each realization, as well as between two intermittent realizations. Finally, the technique is generalized to decrease the probability of error to a significant level. For inkjet-printed PET tags, an evolution of the probability of error is presented in comparison to the optimization of the design of chipless tags. The performance of the system is analyzed by a highly accurate vector network analyzer (VNA)-based reader and a low-cost impulse radio (IR) ultra-wideband (UWB) chipless reader. The probability of error achieved is comparable to the various fingerprint evaluation campaigns found in the literature.

Chapter 1 introduces the chipless RFID technology and its sub-branches. It also discusses the recent developments and advancements in the field of chipless RFID technology. Finally, it presents the challenges of the development of robust authentication techniques.

Chapter 2 presents a brief literature review of numerous existing authentication techniques based on their security level. Apart from existing authentication techniques, this chapter also discusses the necessity of a database for a highly secure authentication application.

Chapter 3 presents a methodology to characterize chipless RFID tags for authentication applications, where procedures to conduct authentication and statistical analyses are presented. The capabilities of chipless technology to be used for tag discrimination are demonstrated using two technologies: PCB and inkjet printing. To validate this approach, three chipless RFID tags are realized. Consecutively from one tag to another, a variation (in the order of fabrication tolerance) is purposely applied to the geometrical dimensions exhibiting the lowest impact on the signal. Chipless tag discrimination based on the level of similarity is presented in both the frequency and time domains.

Chapter 4 presents the extraction of aspect-independent parameters for chipless RFID tags. The extraction of these parameters is needed for authentication because: (i) fewer resources would be needed to save the aspect-independent parameters in the database of authenticity, and (ii) if the chipless tags to be used for authentication are based on multi-scatterers, then the aspect-independent parameters cannot be extracted using only the fast Fourier transform (FFT) approach. Robust detection of depolarizing REP tags using FFT-based short-time Fourier transform is demonstrated. It is demonstrated that, in the frequency-coded chipless RFID technology, as the resonances of the scatterers are orthogonal to each other, the spectrogram method is an efficient and fast choice. The extraction of complex natural frequency(ies) using the spectrogram has never before been performed in the field of frequency-coded chipless RFID. For authentication purposes, aspect-independent parameters are directly associated with the physical dimensions of the scatterer of a chipless tag, but not with the measurement procedure. The random variation of the physical dimensions of the scatterers is then associated with aspect-independent parameters, which is particularly promising for chipless authentication. On the other hand, with an operation of a single measurement, the proposed technique is very promising for the practical implementation of the chipless RFID technology, as it is computationally less expensive due to the inherent fast property of FFT. Thus, the proposed technique requires fewer resources and efforts.

Chapter 5 presents chipless authentication using PCB chipless tags. For this purpose, first, it is shown that the four-coupled C-folded scatterer-based chipless tag is a better choice than the single C-folded scatterer-based chipless tag. Then, the randomness along the geometrical dimensions of a C-folded resonator is analyzed by a second-order bandpass filter model. The concept is proved by fabricating three groups of tags (quad C-folded scatterer tags), which show distinct arms' length, to account for randomness due to the fabrication process. Subsequently, natural dimensional variations in the design of C-folded tags are analyzed for authentication applications. For this purpose, four coupled C-folded scatterer based chipless tags are chosen because of their sharp slope dissimilarity. The chipless tags are realized two times intermittently, where each realization consists of 45 tags. The two different realizations share the same company, the same PCB technology, but a different film mask, in order to ensure the natural dimensional randomness. Similarity analyses are conducted within each realization and between two intermittent realizations. Finally, the technique is generalized to reduce the probability of error to a significant level.

Chapter 6 presents chipless authentication using PET chipless tags printed with a low-cost off-the-shelf available office inkjet printer. The proposed method is based on cheap inkjet-printed square check-patterned tags, whose design is specially optimized by taking the inkjet printing randomness into account. An evolution of the probability of error is also presented in comparison to the vertex-to-vertex gap among the squares of the check pattern. The probability of error achieved is comparable to the various fingerprint evaluation campaigns found in the literature. The performance of the system is analyzed by a highly accurate VNA-based reader and a low-cost IR-UWB reader.

Zeshan ALI
Etienne PERRET
Nicolas BARBOT
Romain SIRAGUSA
June 2020

1

Introduction to Chipless Radio Frequency Identification

1.1. Introduction

In this chapter, we provide an introduction to the chipless RFID technology. After a brief discussion, the recent developments and advancements in the field of chipless RFID technology are presented. In this book, we focus on the development of chipless RFID authentication. For this reason, we also discuss some challenges of the development of robust authentication techniques. This chapter is organized as follows:

– section 1.2 presents the introduction of the chipless RFID technology;

– section 1.3 summarizes the recent developments and advancements from the literature in the field of chipless RFID technology;

– section 1.4 presents numerous challenges of the development of robust authentication techniques;

– section 1.5 concludes this chapter.

1.2. Chipless radio frequency identification

Chipless RFID tags, also called RF barcodes, have several advantages over the conventional passive RFID technology. The absence of any chip (which is the reason it is called chipless) connected to the antenna is the primary revolution of this technology. Chipless RFID is very promising, as it is fully printable, low cost, simple in design and non-line-of-sight operation technology. This technology has enormous potential to replace the barcode in item-level tagging (Perret 2014, Chap. 1).

Coding techniques for the chipless RFID technology can be classified into two main categories: time-coded and frequency-coded chipless tags, as shown in Figure 1.1.

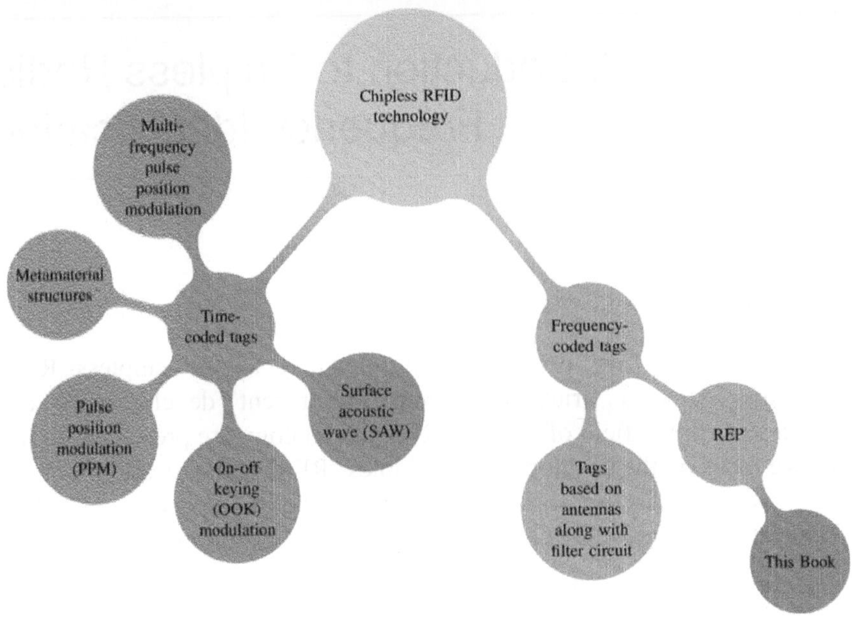

Figure 1.1. *Numerous coding techniques for the chipless RFID technology. For a color version of this figure, see www.iste.co.uk/ali/RFID.zip*

The time-coded chipless technique is first based on sending a pulse signal from the reader to the chipless tag, and then on listening to the backscattered echoes of the transmitted pulse from the tag. The tag code is encoded in the reflected pulse train. On the other hand, in the frequency-coded chipless technique, the tag code is usually encoded by the presence or absence of the peak apexes of resonators. This encoding can also be performed using the phase information at a specified frequency position in the spectrum of the tag. Time-coded chipless tags can be further divided into five categories (Forouzandeh and Karmakar 2015): surface acoustic wave, on–off keying modulation, pulse position modulation, metamaterial structures and multi-frequency pulse position modulation. Frequency-coded chipless tags can be further divided into two categories (Vena *et al.* 2016b, Chap. 4): tags based on dedicated transmission and reception antennas having a filtering circuit between them, and tags based on an RF-encoding particle (REP). An REP is like a scatterer that behaves like a transmitting antenna, a receiving antenna and a filtering circuit simultaneously. The latter technique outperforms the former one in terms of simplicity of design, low cost, low weight and high coding capacity/area. In the former technique, the presence of dedicated transmission and reception antennas causes the mismatching problem, and, ultimately, these antennas do not play their role in increasing the read range. The only advantage of the former technique is that the design of chipless RFID tags shows a separated form.

The radar principle of operation of an REP-based chipless RFID system is schematized in Figure 1.2. A chipless RFID tag is first illuminated by the reader antenna by placing the tag in the field of the reader antenna. The illuminating signal is then coupled with the tag's scatterer. Then, the chipless RFID tag backscatters its response. This backscattered signal is read and stored using the acquisition system.

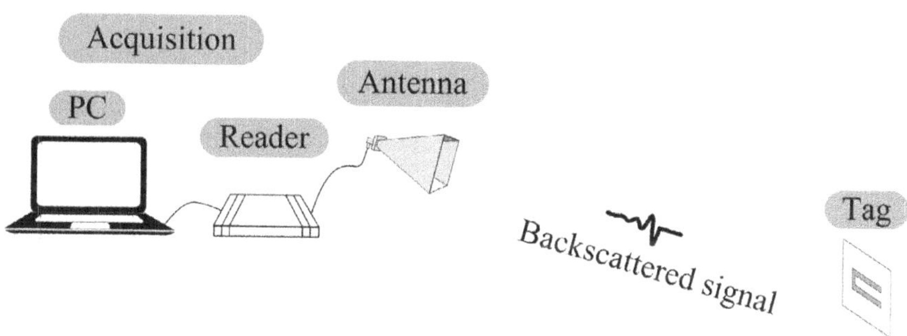

Figure 1.2. *Radar principle of operation of an REP-based chipless RFID system*

Some examples of REP-based chipless RFID tags (Perret 2014, Chap. 5) are shown in Figure 1.3, where REPs are, for example, C-folded scatterer, nested ring resonator, dual-L dipole and shorted 45° dipole. The nested ring resonators and the nested C-folded scatterers provide promising coding density per surface unit, while the nested ring resonators are also invariant to polarization. The dual-L dipole and the shorted 45° dipole provide a depolarizing operation in the illuminated and backscattered waves. On the other hand, a square-shaped scatterer (Betancourt *et al.* 2015) and an octagonal scatterer (Betancourt *et al.* 2016) are also invariant to polarization. Other examples of scatterers are open conical resonators (Nair *et al.* 2014a, 2014b) and quick response (QR) codes such as resonators (Betancourt *et al.* 2017).

In the context of this book, we used REP (e.g. C-folded scatterer, dual-L scatterer, shorted 45° dipole) based chipless tags.

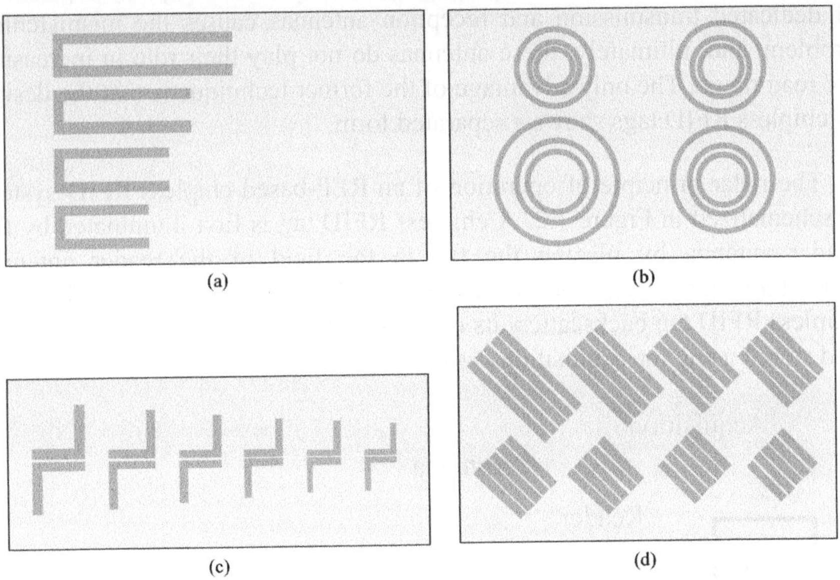

Figure 1.3. *Examples of REP-based chipless RFID tags. (a) C-folded scatterer-based tag. (b) Nested ring resonator-based tag. (c) Dual-L dipole-based tag. (d) Shorted 45° dipole-based tag. For a color version of this figure, see www.iste.co.uk/ali/RFID.zip*

1.3. Recent developments and advancements

Figure 1.4 outlines the recent developments and advancements in the REP-based chipless RFID. Numerous works to enhance the capability of chipless RFID have been reported that are on the aspects of, for example, the tag, the chipless reader, the robustness of detection, sensing and authentication. For the rest of this book, REP-based chipless RFID is simply referred to as chipless RFID.

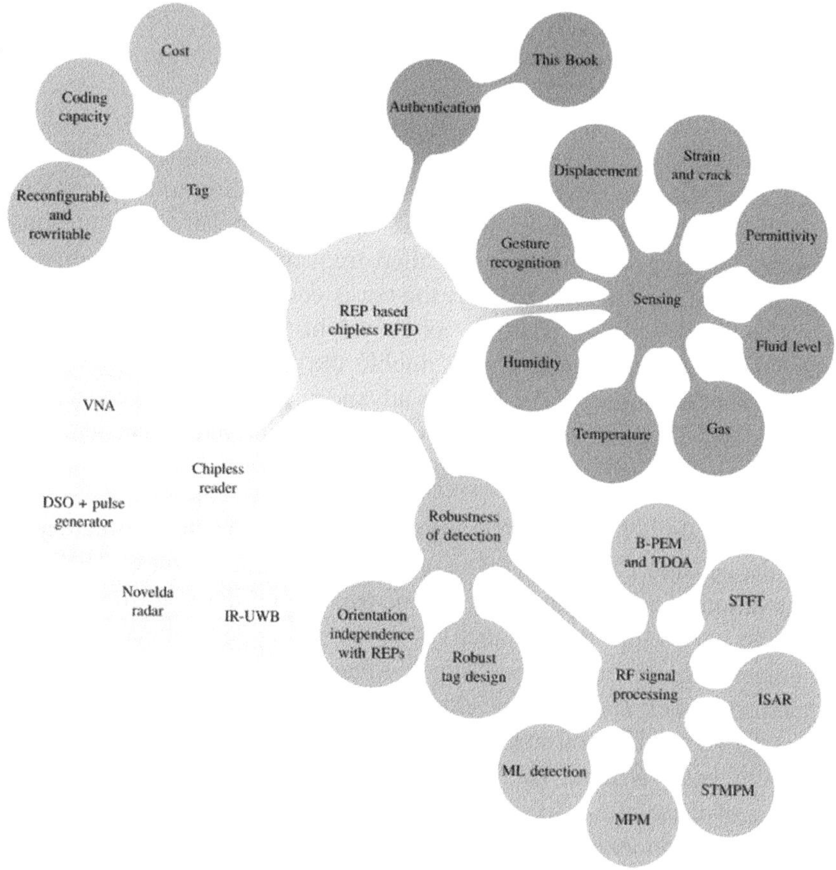

Figure 1.4. *The developments and advancements in the REP-based chipless RFID. For a color version of this figure, see www.iste.co.uk/ali/RFID.zip*

The cost of the chipless RFID has been brought to a few € cents, e.g. €0.4 cents as found in Perret (2014, Chap. 1) and Perret *et al.* (2013), by using the industrial or laboratory equipment. The techniques used are based on:

– printing the paper-based chipless RFID tags using a flexographic technique (Vena *et al.* 2013b);

– printing the PET-based chipless RFID tags using screen printing for fast mass production of tags (Nair *et al.* 2014a, 2014b; Betancourt *et al.* 2015, 2017). Furthermore, a cost reduction of at least 96% or at least 69%

is expected by respectively replacing silver with copper or copper with aluminum with respect to market prices (Barahona *et al.* 2016a).

For improving the coding capacity of chipless RFID tags, the scientific community has intensified its research efforts. Many examples can be found in Khan *et al.* (2016, Table III). Predominantly, encoding in chipless RFID tags is based on the shift of the peak apexes associated with resonant scatterers. This type of encoding is called frequency position encoding. To further enhance the coding capacity, the tag is coded using phase deviations along with the frequency position, as shown in Vena *et al.* (2011, 2016b, Chap. 4). This type of coding may double the coding capacity even with simple REPs (see Figure 1.3). Further advancement of coding capacity has been discussed in Rance *et al.* (2017, Chap. 4), which introduces magnitude coding based on the radar cross section (RCS).

Reconfigurable chipless RFID tags can be divided into two categories: write-only capable chipless RFID tags and rewritable chipless RFID tags. The activation of reconfigurability can be carried out in the form of additive conductive strips on the resonators in an invasive manner (i.e. by a mechanical trigger) or by applying a voltage or laser pulse to specially designed switches (i.e. by an electrical trigger). In write-only capable chipless RFID tags, many non-effective resonators are added in the design of chipless tags. Without the reconfigurability trigger, the frequencies of resonance of these non-effective resonators do not fall within the frequency band of operation of the chipless RFID tag. When the reconfigurability trigger is applied, these additive (non-effective) resonators become effective, showing their frequencies of resonance within the frequency band of operation of the chipless RFID tag. Hence, this category is called write-only capable chipless RFID tags. On the other hand, in rewritable chipless RFID tags, resonators (in the design of chipless tags) are always effective. When the reconfigurability trigger is applied to these effective resonators, there are shifts in the position of the frequencies of resonance within the frequency band of operation of the chipless RFID tag. Therefore, this category is called rewritable chipless RFID tags.

The write-only capable dual-rhombic loop resonators have been presented in Vena *et al.* (2013a). Strictly speaking, this tag shows write-only capability, which is done for the issue of the tag's realization cost.

This tag is first developed by printing the loop resonators with a conductive silver nanoparticle ink and then printing near-transparent strips on the loop resonators using a resistive carbon nanotube ink. By adding resistive strips, the information is written along the amplitude of the RCS level. In addition, the tag provides anti-counterfeiting capabilities due to the near-transparent resistive strips.

The rewritable chipless RFID tags can be reused, because the tags can be rewritten according to the requirement of the user. In Alves *et al.* (2018), a silicon optical switch is used to present a rewritable chipless RFID tag. The optical switch can change its state when illuminated by a laser source. This concept is proved by using a filter-like configuration that does not fall within the category of REP-based chipless RFID. However, the same concept can also be applied to REPs. Furthermore, this rewritable chipless RFID tag can only maintain the reconfigurability effect in the presence of the laser pulse. Therefore, this proposed rewritable chipless RFID tag is not bistable.

In Jayakrishnan *et al.* (2020), a rewritable chipless tag is presented. This rewritable scatterer is based on a non-volatile memory switch which, in turn, is based on the conductive-bridging random access memory (CBRAM) or metal–insulator–metal switch. The two examples of the reconfigurable scatterer discussed above show a separation of 200 MHz and 140 MHz between the on and off states of the CBRAM switch, respectively. Such values of separation seem to be sufficient to distinguish between the on and off states of the CBRAM switch to develop a rewritable chipless RFID tag. Furthermore, this rewritable chipless RFID tag can maintain the reconfigurability effect even in the absence of the voltage pulse. Therefore, this proposed rewritable chipless RFID tag is bistable.

For laboratory experiments, the most commonly used equipment for chipless RFID readers in the scientific community is based on:

– the vector network analyzer (VNA);

– the digital sampling oscilloscope along with a pulse generator.

This high-cost equipment (several tens of thousands of euros) is not feasible for the practical implementation of the chipless RFID technology for item-level tagging. One solution is to use a Novelda NVA-R640[1]

1 See https://www.xethru.com/.

development kit radar as a chipless RFID reader, as discussed in Vena *et al.* (2011, 2015a) and Barahona *et al.* (2017). The cost of the Novelda NVA-R640 development kit radar is €2000, including the low-noise amplifier that is still expensive for low-cost applications. In addition, the Novelda radar shows bandwidth limitations. The 3 dB achievable bandwidth is 4.5 GHz, ranging from 1.5–6 GHz. Therefore, the resonant scatterers in the design of chipless tags operating above a frequency of resonance of 6 GHz cannot be detected precisely using the Novelda radar. To overcome the cost and bandwidth limitations, an impulse radio (IR) ultra-wideband (UWB) reader has been proposed in Garbati *et al.* (2015, 2017). This IR-UWB-based reader is developed using off-the-shelf available components that result in a low-cost device. The cost of this reader is approximately €1000. The band of operation of this reader is 3.1–10.6 GHz.

For the robust detection of chipless RFID tags, two techniques have been discussed in the literature: robust tag design and RF signal processing. A robust tag design is required for the following reasons:

– currently, the majority of the tags' designs are uniplanar to make them fully printable. In this case, peak apexes associated with resonant scatterers can show random shifts due to random changes in permittivity that occur due to the absence of the ground plane;

– disorientation in the reading process can induce random shifts in peak apexes associated with resonant scatterers.

In Vena *et al.* (2012a, 2012b), frequency shifts are compensated by equipping one resonator as a constant, which means its geometry is the same for all the tags and thus cannot be used for identification.

For the orientation of reading independence, numerous REPs have been proposed in the literature: for example, the nested circular ring resonator (Islam *et al.* 2012; Vena *et al.* 2012c), the dual-L depolarizing scatterer and the shorted 45° depolarizing scatterer (Vena *et al.* 2013c), the nested cross loop resonator (Sajitha *et al.* 2016), the square-shaped scatterer (Betancourt *et al.* 2015) and the octagonal scatterer (Betancourt *et al.* 2016). In Garbati *et al.* (2016), an orientation-independent reading system has been proposed, where the transmitted signal can be rotated electrically with fixed antennas to preserve the cross-polarization for depolarizing chipless tags.

For the robust detection of chipless RFID tags, robust RF signal processing techniques are required for multiple reasons:

– a misalignment between the reader and the tag can lead to a decoding error;

– if the tag is placed on an irregular or asymmetrical surface, the random superposition of the structural mode and the tag mode can create shifts in resonant peak apexes that can again lead to a decoding error;

– until now, most of the decoding techniques for chipless RFID require two measurements: an empty measurement (i.e. a measurement in the absence of the tag) and a tag measurement (i.e. a measurement in the presence of the tag). If the tag is affixed to an object, then the empty measurement is impossible, which finally leads to a decoding error.

For the characterization of tag identification (ID) in chipless RFID tags, an Euclidean distance-based minimum distance detecting method (Barahona *et al.* 2016c) and a maximum likelihood method have been discussed in the literature. On the other hand, to decode chipless RFID tags, complex natural resonances (CNRs) are extracted using the matrix pencil method (MPM) (Blischak and Manteghi 2011) and its variant short-time matrix pencil method (STMPM) (Rezaiesarlak and Manteghi 2013). In Rezaiesarlak and Manteghi (2014a), the STMPM is applied to extract the high-dense tag code. In Rezaiesarlak and Manteghi (2015), the early-time and late-time modes of the transient response from multi-scatterer targets have been distinguished using the STMPM. In Costa *et al.* (2018a), the tag code of mobile chipless RFID tags has been extracted using inverse synthetic aperture radar data processing. However, these techniques are based on two measurements: an empty and a tag measurement. In Ramos *et al.* (2016), tag detection for the chipless RFID technology in different environments using a technique based on the short-time Fourier transform (STFT) is addressed. With this technique, the tag ID is extracted without background normalization (i.e. single measurement) by using an averaged late-time signal. It has been shown that the technique is efficient even without background normalization operating in a realistic outdoor environment. Hence, this technique is very promising for the practical implementation of chipless RFID. For the decoding the same-coded tags, different techniques corresponding to the tag placement have been presented in Barahona *et al.* (2014, 2016b). For decoding the line of sight and same-coded in-line placed tags in the reader zone, a backscattered pulse energy modulation scheme (i.e. based on the changes in the received RCS level) has been proposed in Barahona *et al.* (2014). For decoding the chipless tags at different distances in the reader zone, the time difference of arrival of backscattered signals has been exploited in Barahona *et al.* (2016b).

Apart from the identification applications, wireless sensing capabilities of the chipless RFID technology have also been discussed in the literature. Such additional sensing capabilities of the chipless RFID technology are very beneficial for environmental monitoring and industrial control. Sensing features in chipless RFID tags are based on the shift of the peak apexes associated with resonant scatterers in most of the studies reported in the literature.

For humidity sensing, in Amin *et al.* (2014), a chipless RFID tag has been proposed for identification and relative humidity (RH) sensing, where a patch loaded with multiple slots is used for the tag code and a single electric inductive–capacitive resonator on a polyvinyl alcohol film is used for RH sensing. In Feng *et al.* (2015), a paper-based chipless RFID tag designed with inductor–capacitor resonators has been presented for humidity sensing. In Borgese *et al.* (2017), humidity sensing has been proposed using a frequency-selective surface (FSS) (i.e. consisting of three concentric loops) based inkjet-printed chipless RFID tag. In Vena *et al.* (2016a), a chipless RFID tag for identification and RH sensing applications has been proposed, where a multiple coupled loop resonator is used for the tag code and a deposited layer of silicon nanowires is used for RH sensing. A similar concept has been presented in Deng *et al.* (2018), where a slotted patch is used for the tag code and a deposited layer of silicon nanowires is used for RH sensing.

For the temperature as well as CO_2 sensing, in Vena *et al.* (2015b), a split-ring resonator (SRR) based inkjet-printed chipless RFID tag with three different inks has been proposed. In the design of this chipless sensor, a deposited layer of a composite polymer/single-walled carbon nanotube ink is used as a transducer. For temperature sensing, in Matbouly *et al.* (2018), a chipless RFID tag compliant with RF emission regulations has been proposed for temperature sensing, where a Rogers RT/Duroid 6010.0LM dielectric substrate is used as a transducer. The tag is based on a C-folded scatterer with embedded slots operating only in allowed bands: European Telecommunications Standards Institute (ETSI) RFID band and Industrial, Scientific and Medical (ISM) 2.5 GHz and ISM 5.8 GHz bands. In Lu *et al.* (2018), a high-temperature chipless RFID tag based on a gold (Au) microstrip slotted patch has been proposed, where an alumina substrate is used as a transducer. The working range of this proposed sensor is 25–800°C with an average sensor sensitivity of 101.94 kHz/°C.

For the detection of fluid level, in Guillet *et al.* (2012), coplanar stripline (Garg *et al.* 2013, Chap. 7) based C-folded scatterers have been used to determine the water level. In this system, C-folded scatterers are pasted on a water container and the level of water is determined by the diminishing resonant peak apexes by filling the water tank step by step. This technique is very promising because the low-cost C-folded scatterers can be printed on the container during the production process.

For the estimation of the permittivity of different materials, in Perret (2016), a chipless RFID technique based on RCS measurements has been proposed for the first time. Similarly, in Costa *et al.* (2018b), two 45° dipole-based chipless RFID tags have been used to estimate the permittivity of different materials. This proposed technique can also be used for monitoring changes in the electrical properties over time. In Lázaro *et al.* (2018), a chipless RFID tag based on an FSS loaded with printed capacitors has been proposed to estimate the permittivity of the material to which the tag is attached. The main application of this chipless sensor tag is to monitor civil structural health.

For strain and crack sensing, in Vena *et al.* (2014a, 2014b), an SRR-based inkjet-printed chipless RFID tag on a polyimide substrate has been proposed. In this chipless sensor, deformations in printed strips due to applied strain produce amplitude variations in the RCS. This variability is used for strain and crack sensing. In Marindra and Tian (2018), a chipless RFID tag has been proposed to detect and characterize cracks in metallic structures. This sensor tag is based on four dipoles along with a circular microstrip patch antenna (CMPA) resonator, where the orientation and width of cracks can be detected using the behavior of shifts in the peak apexes associated with resonators.

For sub-millimeter displacement sensing, in Perret (2017), the phase of the backscattered signal from a chipless RFID tag has been exploited. This proposed system has shown that displacements of the object can be detected using chipless RFID with a possible resolution of less than 1 mm, even in a realistic outdoor environment with the surrounding objects or obscured by opaque objects. This feature of displacement sensing has been added to the predesigned chipless RFID tag for identification applications (see Figure 1.3(c) and (d)) without compromising the coding capacity.

For gesture recognition, in Barbot and Perret (2017), a human–computer interaction system has been proposed to detect and localize a human finger on a chipless RFID tag. In this system, a dielectric paste is placed on different scatterers on a chipless RFID tag, and the position of the dielectric paste can be detected by the presence or absence of the peak apexes associated with the resonant scatterer. This feature of gesture recognition has been added to the predesigned chipless RFID tag for identification applications (see Figure 1.3(d)) without compromising the coding capacity.

For 2D localization sensing, in Hu *et al.* (2010), a chipless RFID tag based on a coplanar waveguide-fed monopole antenna has been proposed. Then, tag detection and localization have been demonstrated using a differential delay-and-integrate receiver. In Anee and Karmakar (2013), three different antennas have been used to localize the chipless tag by analyzing the early-time response (structural mode) using the trilateration algorithm. In Rezaiesarlak and Manteghi (2014b), a cellular technique has been used, where each cell has a triangular geometry with three antennas at the vertices. Then, chipless tags are localized by calculating the roundtrip time by applying the narrow-frequency matrix pencil method to the early-time response (the structural mode in the frequency domain). In Barbot and Perret (2018), the phase offset of the backscattered signal from a known position to an unknown position has been exploited to localize a chipless RFID tag (or the object to which the tag is attached) on a 2D plane. In this system, a simple search method (multilateration algorithm) is used to localize the chipless tag with an accuracy of less than 4 mm. This feature of localization is added to the predesigned chipless RFID tag for identification applications (see Figure 1.3(d)) without compromising the coding capacity.

Finally, this book discusses a novel aspect of the chipless RFID technology that is extended to the chipless authentication. The basic idea of the proposed chipless authentication can be explained by the arrangement of paper fibers. Figure 1.5 shows the scanning electron microscope (SEM) photograph of an ordinary paper. The unique arrangement of paper fibers naturally occurs during the realization process. Copying such a naturally occurring pattern in an exact manner is virtually impossible. Similarly, Figure 1.6 presents an overview of the chipless authentication concept. Any two RFID tags developed using the same digital design (e.g. film mask for printed circuit board (PCB) chipless tags) will show process variations in their geometrical dimensions. These variations can be:

– the non-systematic over- or under-etching in the case of the PCB;

– the randomness of ink drops in the case of inkjet printing.

These independent variations can produce unique signatures that can be used for authentication. A comparison among the signals from the same device will produce a theoretical value of similarity level equal to 1. Moreover, a comparison between two different devices will produce a theoretical value of similarity level equal to 0.

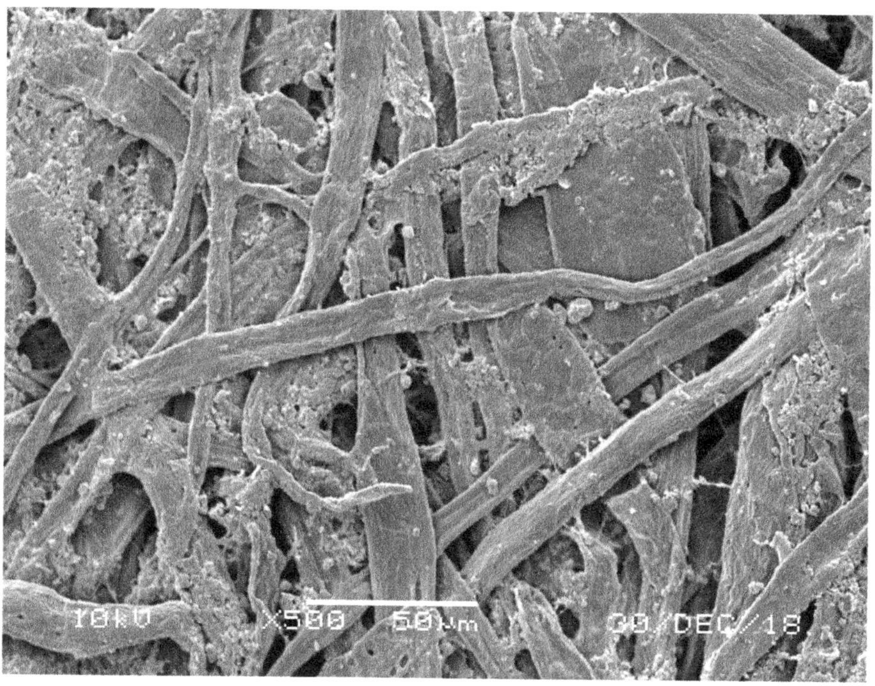

Figure 1.5. *SEM photograph of a normal paper. For a color version of this figure, see www.iste.co.uk/ali/RFID.zip*

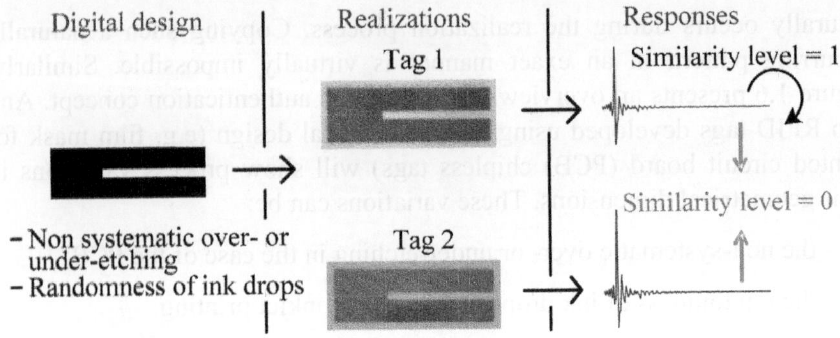

Figure 1.6. *Overview of the chipless authentication concept*

1.4. Authentication

Anti-counterfeiting and authentication techniques are widely used in the industry to evaluate the authenticity of the products. Various techniques ranging from less secure visible markers (e.g. labels) to highly secure biological elements (e.g. DNA – deoxyribonucleic acid) are currently being used. The robustness of an authentication technique can be defined as follows:

- highly secure;
- non-invasive or non-intrusive operation;
- difficult to duplicate;
- simple in operation;
- low cost;
- ease of fabrication.

The development of robust authentication techniques remains a challenge because of the numerous requirements mentioned above.

This book is focused on taking the next step with the aim of developing chipless tags for authentication applications. The concept of chipless RFID is extended to authentication where each tag has to present a unique signature that can never be reproduced even if someone tries to copy the tag.

For this purpose, natural randomness (i.e. inherent in the fabrication process) along the dimensional parameters of resonators is used. Such natural randomness can produce unique electromagnetic (EM) signatures that can be used for authentication.

1.5. Conclusion

In this chapter, the chipless RFID technology and its sub-branches were briefly explained. In addition, the recent developments and advancements from the literature in the field of chipless RFID technology were summarized. Finally, the challenges of the development of robust authentication techniques were discussed.

2

Literature Review

2.1. Introduction

Counterfeiting has become a global and dynamic phenomenon because it is complicated to measure the total inter-border trade of counterfeit items and the global worth of this illegal industry. The available estimates of the value of the counterfeit trade are based on the seizures by law enforcement agencies. According to the US government estimation in 2008, the counterfeit industry was $500 billion worth in the global market. This illegal industry had a 1700% growth rate over a time span of 10 years from 2008 to 2018 (Chaudhry and Zimmerman 2013, Chap. 2). With an extreme estimation, the total international trade of counterfeited items is in the range of 5–7% of the global trade. On the other hand, the estimation of the volume of inter-border trade of counterfeit products was $250 billion (OECD 2009). According to the OECD and EUIPO (2016) in 2013, the total international trade of counterfeited items was up to 2.5% (i.e. $461 billion) of the global trade. Based on the value presented in OECD and EUIPO (2016), the International Chamber of Commerce has projected the possible value of counterfeit goods from $1.90 to $2.81 trillion in 2022. Such huge black money can play a negative role to disturb the harmony of the world (Obama 2011). In addition, this illegal practice poses a threat to a wide range of industries and harms societies from various perspectives:

– ultraexpensive consumer goods (e.g. cosmetics, fragrances, leather articles, jewelry);

– business-to-business industrial goods (e.g. tools, appliances, materials, medical devices, replacement parts of automobiles and machinery);

– essential consumer goods (e.g. food items and medicines).

In this chapter, different proposed techniques from the literature are collected and discussed. This chapter is organized as follows:

– section 2.2 presents numerous authentication techniques (by category first), then further sections discuss each of the techniques from the selective literature;

– section 2.3 concludes this chapter.

2.2. State of the art

Security technologies offer vast opportunities to combat fakes in the global supply chain. According to Seena (2014), security technologies can be divided into two classes, as shown in Figure 2.1: authentication solutions and serialization. Serialization (which is further subdivided into identification and track and trace methods) is based on first applying the unique identifiers to the products and then keeping the records of the products from one stage to another using distributed networks. In this method, each activity of the product is broadcasted in the networks to update all the nodes in order to counter the fake items. On the other hand, authentication solutions are subdivided into two classes: logical authentication and physical authentication. Logical authentication is mostly related to counter the piracy of digital media, for example, software distributions, images, digital paintings and films. In this book, we are interested in the physical authentication methods. Generally, these solutions can be categorized into three types (Power 2008; Li 2013): visible features or overt, hidden markers or covert, and forensic methods. Figure 2.2 explains the existing anti-counterfeiting or authentication techniques that are categorized based on the security level. The security increases as we move from level 1 to level 3. In terms of applications, there is a broad spectrum of solutions ranging from simple solutions with low security level (generally no equipment is needed to detect the presence of the security tools) to more complex solutions with high security level (based on the use of specific electronic equipment and the use of the random aspect of the medium).

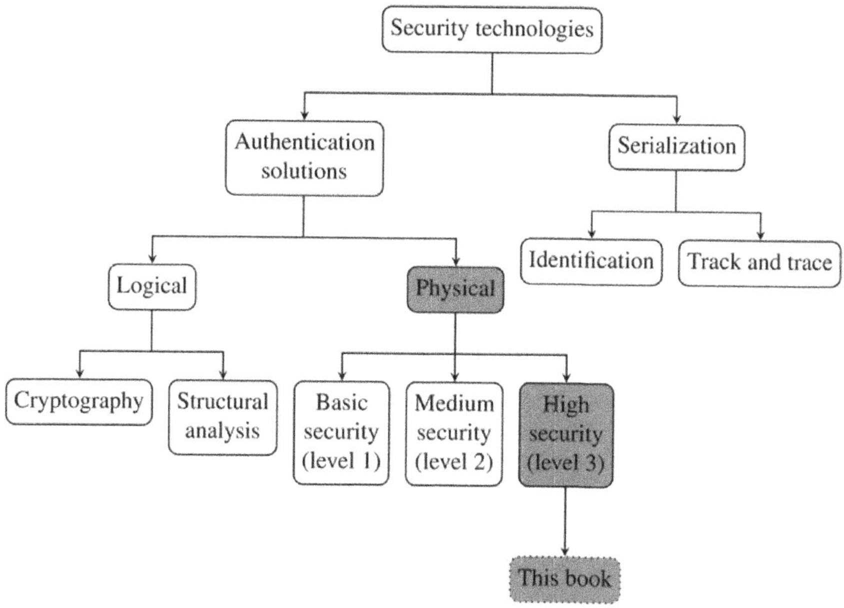

Figure 2.1. *Types of security technologies. For a color version of this figure, see www.iste.co.uk/ali/RFID.zip*

For security levels 1 and 2, a database of authenticity is not required and authentication can be performed by observing specific features of the authentication elements. To elevate the security level of an authentication technique, a database of authenticity can be used, which necessitates a comparison between the measurements at the instant of authentication and the measurements stored in the database. For this reason, a database in security level 3 is essential for the highly secure application. In this book, we focus on presenting a highly secure (security level 3) physical product authentication method (see green blocks in Figure 2.1) by extending the capability of the chipless RFID technology.

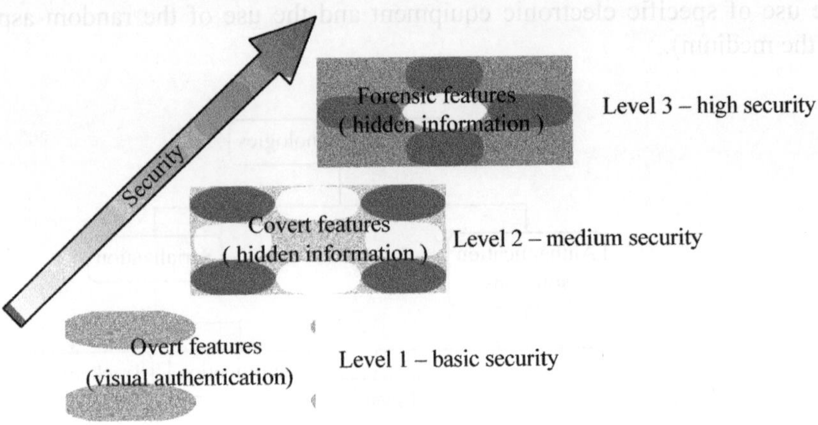

Figure 2.2. *Existing product anti-counterfeiting or authentication techniques categorized based on the security level. For a color version of this figure, see www.iste.co.uk/ali/RFID.zip*

2.2.1. *Basic level of security (overt or visible features)*

Overt features are the first-line features (security level 1) as they provide the basic security level. These are the visible authentication features that can be detected and analyzed by naked eyes without the use of any special tools. These features are mainly designed for the untrained common consumer to verify the authenticity of purchased items (Glossary Terms IACC). Apart from this, these features can also be used by goods suppliers, distributors, food and drug administration, representatives of pharmaceutical companies, customs inspectors, law enforcement organizations and the court of justice. Some examples of the overt or visible features are:

– holograms (Mallik 1992; McGrew 1995; Pizzanelli 1997);

– watermarks (Huang and Wu 2002, 2007);

– optically variable devices (Berning and Phillips 1990; Wild and Brehm 2005);

– color-shifting inks and films (Bradley and Witzman 2001);

– security graphics (Moore 1997);

– security thread (Lawandy 1998; Seifert 1999);

– intaglio printing (or raised printing) (Hutton and Merry 1977; Nemeth 2002);

– sequential product numbering (Brock 2001; Greer and Wiklof 2002);

– on-product marking (Vig and Saglimbeni 2005; Smith *et al.* 2012);

– scratch-off layer on barcode (Royer 2001).

Figure 2.3 shows some reproducible examples of visible features (security level 1). These examples are generated using a €50 currency bill. Figure 2.3(a) shows a portrait watermark of Europa. Figure 2.3(b) shows a security thread with an embedded symbol of € sign and a numeral 50. Figure 2.3(c) shows a transparent hologram. This hologram becomes transparent when seen against the light. Otherwise, this hologram presents a numeral 50 along with the rainbow pattern and a portrait of Europa from different angles. Figure 2.3(d) shows a raised print (intaglio print) that can be observed by its tactile effect. Figure 2.3(e) shows a numeral 50 printed with a color-shifting ink. This numeral changes its color from deep green to deep blue when viewed at different angles. Figure 2.3(f) shows a latent image on the rear side of the €50 currency bill. This strip is only visible at a certain angle.

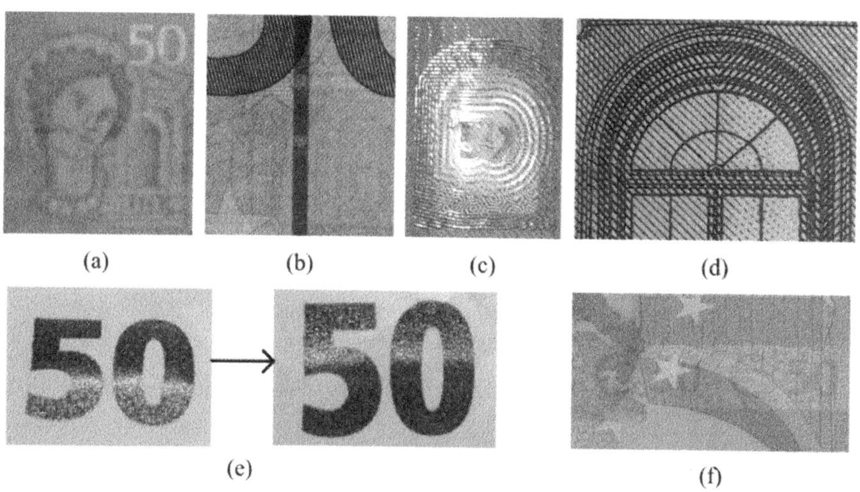

Figure 2.3. *Some examples of visible features (security level 1) generated using a €50 currency bill. (a) Watermark. (b) Security thread. (c) Transparent hologram. (d) Intaglio printing. (e) Color-shifting ink. (f) Partial latent image. For a color version of this figure, see www.iste.co.uk/ali/RFID.zip*

2.2.2. *Medium level of security (covert or hidden markers)*

Covert features are the second-line features (security level 2) as they provide a medium level of security. These are the hidden authentication markers that cannot be detected and analyzed by the naked eye. These features are not designed for the common consumer. Instead, special tools (or assistance) are required to investigate them. These features are for goods suppliers, distributors, food and drug administration, representatives of pharmaceutical companies, customs inspectors, law enforcement organizations (Glossary Terms IACC). Apart from this, these features can also be used by the court of justice. Some examples of the covert or hidden features are:

– invisible printing (Outwater and Loop 2001; Outwater and Helmick 2002; Cruikshank *et al.* 2006);

– magnetic ink (Raksha *et al.* 2015);

– fluorescent inks (Auslander and Berson 1997; Tan 1997);

– micro- and nano-marks (Phillips 2004; Fragala *et al.* 2011);

– optical methods (Carro-Temboury *et al.* 2018);

– embedded images (Koltai *et al.* 2000; Golan 2008);

– digital watermarks (Lawandy and Tillotson 2001; Rhoads 2013);

– hidden (or latent) images, marks and printing (McCarthy and Swiegers 2007);

– glyphs (Cass and Marimont 1997; Mazaika 2004);

– anti-copy or anti-scan design (Gardner and Voticky 1985; Phillips and Phillips 2005);

– laser codes or marks (Levy *et al.* 2005);

– security paper (Howland and Foulkes 1999; Foster and Mulcahy 2000);

– substrates;

– odor.

Figure 2.4 shows some reproducible examples of covert features (security level 2). These examples are also generated using a €50 currency bill. Figure 2.4(a) shows micro-prints at various positions of the front side of the €50 currency bill. At present, the end consumer can also analyze these micro-prints due to the proliferation of smart mobile phones with high-resolution cameras. Figure 2.4(b) shows some fluorescence features under ultraviolet light. On the rear side, the latent image turns into pink color and one-fourth of the circle turns into yellow color. On the front side, the embedded small circles and the stars turn into yellow color. Apart from these hidden features, embedded fibers glow in red, blue and green colors.

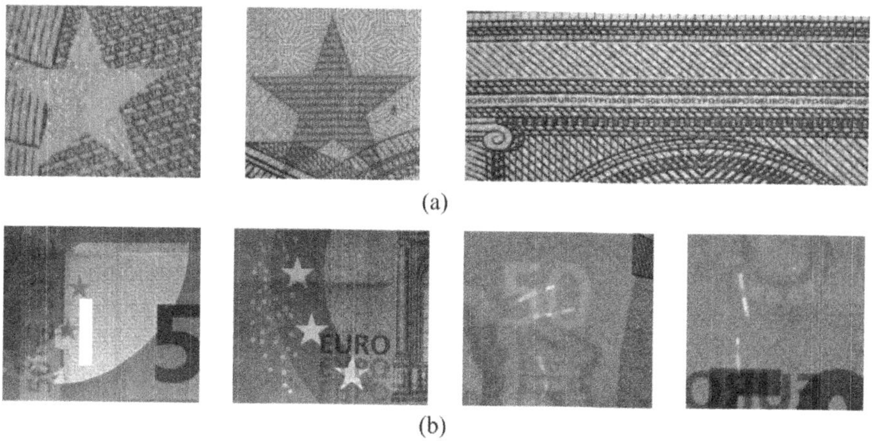

Figure 2.4. *Some examples of covert features (security level 2) generated by using a €50 currency bill. (a) Micro-prints. (b) Fluorescence features under ultraviolet light. For a color version of this figure, see www.iste.co.uk/ali/RFID.zip*

2.2.3. *High level of security (forensic techniques)*

Highly secured features (security level 3) provide more complex solutions with high security level (based on the use of specific electronic equipment and the use of the random aspect of the medium). Most of these features can only be analyzed using forensic examination in high-tech laboratories (Glossary Terms IACC).

These techniques are, for example, based on (Gooch *et al.* 2016):

– chemical taggants (Gaynor *et al.* 2018, 2017);

– biological taggants (Eastwood *et al.* 2011);

– DNA taggants (Sheu *et al.* 2002, 2006; Mercolino 2013);

– isotope taggants (Welle 1998; Frideling 2006);

– micro-taggants (Loving 1995; Brogger *et al.* 2004; Stierman *et al.* 2008);

– spectroscopic taggants (Kerns *et al.* 2006; Lawandy *et al.* 2014).

2.2.4. Conventional RFID approaches

Conventional passive RFID refers to a wireless system that is based on two devices in its fundamental form: a reader and a tag. The tag consists of an IC (integrated circuit) chip and antenna (Rao *et al.* 2005). The purpose of the former component is, for example, to store the tag ID and to implement the communication protocol. The latter component is used for the RF communication between the tag and the reader. The reader communicates with the tag by sending a composite signal to the tag. This composite consists of a modulated signal and pulses of an unmodulated signal. After receiving the reader's signal, the tag modulates the unmodulated signal and backscatters it to the reader. The tag can exhibit a read range from a few millimeters to several meters. However, the size of the tag antenna can become large for an extended range RFID tag, which will ultimately increase the size of the tag.

Initially, the research efforts for the development of the conventional passive RFID technology were focused on identification as its primary application. With the advancement of research, various other applications of the RFID technology have been explored. These applications, primarily in the field of sensing, for example, are humidity sensing (Feng *et al.* 2015), temperature sensing (Vena *et al.* 2014a, 2014b; Zannas *et al.* 2018), light sensing (Salmerón *et al.* 2014a, 2014b), displacement sensing (Paggi *et al.* 2014), pressure sensing (Rennane *et al.* 2018) and crack monitoring (Donelli 2018).

Apart from the above-discussed applications, the RFID technology is also applied for authentication solutions. Strictly speaking, an RFID tag cannot be directly used for authentication because it is susceptible to be copied easily. Authentication differs from identification in that authentication identifies and determines the genuineness of its subject. One way to use the RFID technology for verifying the authenticity of the products is to use digital solutions such as serialization/track and trace. The other two methods are RF fingerprinting and near-field physical-layer identification. Both approaches use the analog features of backscattered signals.

2.2.4.1. *Serialization/track and trace*

Serialization corresponds to assigning a unique identifier to each product individually or to a batch of products. This unique identifier can be, for example, alphanumerics, barcodes, QR codes or a code stored in an RFID tag.

The track and trace technology corresponds to a process of monitoring and recording all the movements of a product in a supply chain. This monitoring and recording is usually carried out using a distributed network. Tracing corresponds to keeping the records of the intermediate destinations of the product (whereabouts), while tracking refers to keeping the record of the next intermediate destination (Glossary Terms IACC).

The overall serialization/track and trace solution (digital solution) is based on first applying the unique identifiers to the products and then keeping the records of the products from one stage to another using distributed networks. Each activity of the product is broadcasted in the networks to update all the nodes in order to counter the fake items. Some examples of product serialization/track and trace are unique serial numbering, barcodes, RFID tagging, and track and trace-based plausibility check (Lehtonen *et al.* 2008, Chap. 9; Li 2013).

In the context of conventional passive RFID, various track and trace methods have been proposed in the literature. An anti-counterfeiting system and a tag data processing and synchronization (TDPS) algorithm to track and trace using RFID have been discussed in Choi *et al.* (2015). This TDPS algorithm produces an initial secure electronic pedigree (e-pedigree) for clothing merchandise during production. This method further keeps the records of the whereabouts of products from industry to dealers, in order to provide for product authentication. Another track and trace method for RFID-enabled supply chains has been proposed in Zanetti *et al.* (2010). This method takes into account tag misreads that do not depend on the global knowledge of the supply chain. Furthermore, ad hoc protocols have been implemented and tested for this proposed track and trace system. On the other hand, a highly accurate approach to detect duplicate tags in distributed RFID systems has been proposed in Kamaludin *et al.* (2018) with a detection accuracy of 99%. This proposed scheme is based on the consistency of dual hash collisions and a modified count-min sketch vector (i.e. a data structure based on dual independent hash functions). However, even with the high performance of the track and trace algorithms, in the post supply chain, RFID tags are prone to be duplicated or cloned. To solve this security loophole for the post supply chain, Bitcoin's blockchain-based product ownership management system (POMS) has been proposed in Toyoda *et al.* (2017). This Bitcoin's blockchain decentralized system provides an opportunity to the customer to verify the evidence of custody of merchandise with a cost of less than $1 to manage the ownership of a product with up to six transfers.

2.2.4.2. *RF fingerprinting*

RF fingerprinting is the second way to use conventional RFID tags in authentication applications. This methodology is based on using the analog features of RFID tags. In this context, an electronic fingerprinting method for RFID tags based on their physical attributes has been proposed in Periaswamy *et al.* (2011). The physical attributes are extracted using the minimum power responses of the tags measured at several frequencies. On the other hand, these attributes are extracted using the dynamic wavelet fingerprint, which are used to authenticate several individual RFID tags (Bertoncini *et al.* 2012). Here, for the classification of unique RFID tags, a supervised pattern classification technique is used. In another work (Yang *et al.* 2015), the phase information of the backscatter signals of RFID tags is used to create hardware fingerprints. For the authentication of electronic chips (Kheir *et al.* 2014), first, micro- and nanostructured composite materials mixed with a dielectric fixing matrix are proposed. Then, fingerprints are characterized using statistical similarity measures such as correlation coefficient and Tanimoto distance.

2.2.4.3. *Near-field physical-layer identification*

As a third approach of authentication based on conventional RFID tags, the near-field physical-layer identification is also exploiting the analog features of RFID tags. Here, the fingerprints that have been exploited for authentication, in the literature, are based on the magnitude and phase information of chosen key frequencies (3rd and 5th harmonics) of backscattered signals (Romero *et al.* 2009), the modulation shape and spectral features of the response signals of backscattered signals (Danev *et al.* 2009) and the normalized power spectral density of the envelope of backscattered signals. Another feature extraction technique for high-frequency (HF) RFID cards has been presented in Danev *et al.* (2012), where principal components analysis is one of the commonly used feature selection metrics. Subsequently, in Romero *et al.* (2010), identification of individual HF RFID cards of the same type is presented, where the measured features are the unloaded resonance frequency, quality factor or tag response to a special signal.

It is important to mention that it is difficult to design a reliable authentication system using the classical passive RFID tags because all chips (of RFID tags) may exhibit the same type of backscattered signals. Hence, it may be complicated to separate the part of the signal coming from the tag and the part coming from the environment. Finally, the classical passive RFID tags can also be cloned easily. Therefore, chipless RFID is a better choice to be used in authentication applications. The backscattered response from a chipless RFID tag is analog in nature, which also represents the physical layer of communication.

2.2.5. *Classical chipless approaches*

Chipless RFID has also been used in authentication in the literature, where the elements of the authentication are in the form of dipoles (Greene and Hurley 2002) or a plurality of conductive wires (Marchand 2003). A hybrid (identification/authentication) chipless RFID has been proposed in Perret *et al.* (2012). In this system, the chipless RFID technology has partially been used as an authentication tool, whereas the optical technology is used for identification applications. For secure proximate field data transfer, particularly for the identification of bank cards, an RFID system has been presented in Deepu *et al.* (2010), where a cavity resonator is used to differentiate chipless tags by exploiting the shifts in the frequency of resonant modes. Another proximate (near field) 40-bit split-ring resonators (SRRs) based chipless tag along with an SRR-loaded microstrip line reader have been introduced in Herrojo *et al.* (2017). The underlying possible applications of this system discussed are the security and authentication applications. RF absorbing markers (based on particles of inorganic oxide materials) have also been used for authentication purposes such as security documents (Romero *et al.* 2017). Conversely, a tagless RFID approach for identification of electronic gadgets has been presented in Yang and Sample (2016), where EM emissions from all electronic appliances are first captured and then compared using the cosine similarity function as a resemblance benchmark.

2.2.6. *Natural randomness*

Natural randomness during the fabrication process of chipless RFID tags can be used to elevate the security level of authentication systems. In such a case, chipless tags would then be very difficult to duplicate. An intuitive example of natural randomness is presented in Figure 1.4, with the help of a zoom photograph of paper fibers. Such unique patterns have been exploited for authentication in Boutant *et al.* (2007), where the natural texture of the surface of the product (e.g. fiber arrangement of the paper) is first converted into a digital signature and then used for authentication.

In the recent past, an unclonable chipless RFID tag in the same concept has been presented in Yang *et al.* (2016), where natural process variations of slot parameters (trace width, air gap, thickness of the substrate and dielectric constant of the substrate) are assumed and the Euclidean distance is used to quantify the uniqueness of tags. An alternative hardware-based method to create RF fingerprints by exploiting the proximate RF effects between several antennas of the RFID reader and the randomly modified substrate of RFID tags has been proposed in Lakafosis *et al.* (2011). In addition, such a hardware-based fingerprinting method is also proposed for a flexible substrate (DeJean *et al.* 2011). A concept of RF certificates of authenticity (RF-CoA) for authentication has been discussed in DeJean and Kirovski (2007). These RF-CoA are realized with a random arrangement of thin conductive plates of arbitrary planar shape. Otherwise, for authentication in the printing industry, a plurality of resonators with a random arrangement within the base material (e.g. paper, ink, coatings, polymers, composites, adhesives) has been presented in Christofferson *et al.* (2009).

The techniques of security levels 1 and 2 (overt and covert features) might easily be mimicked. In security level 3, forensic techniques may be invasive or destructive in operation. In addition, the necessity of using laboratory equipment is a drawback. It is a costly solution that may not be possible to implement low-cost items for authentication. Serialization/track and trace with barcodes require line-of-sight operation. Even for RFID labeling, this technique requires a common database structure and common standards across the markets. The authentication techniques based on chip-based RFID (Danev *et al.* 2009, 2012; Romero *et al.* 2009, 2010; Periaswamy *et al.* 2011; Bertoncini *et al.* 2012; Kheir *et al.* 2014; Yang *et al.* 2015; Zhang *et al.* 2016) can be bulky or costly due to the use of the chip. In the category of natural randomness-based chipless approaches, the techniques based on RF-CoA (DeJean and Kirovski 2007; DeJean *et al.* 2011; Lakafosis *et al.* 2011) may require 3D structures. The technique presented in Christofferson *et al.* (2009) may require a particular printing procedure. The optical technique (Boutant *et al.* 2007) may have a limitation of the close imaging procedure, which can also be imitated.

In comparison with the state of the art, we propose for the first time that the realizations of simple scatterers can be used as unique footprints. The proposed approach is very low cost and simple because we do not add any particle to the labels, as done in Perret *et al.* (2012). The natural dimensional variations in the realizations of chipless tags give rise to the unique variations in the RF field that is compatible with high-level security. An in-depth statistical analysis is presented in Chapter 3 to provide an accurate idea of the performance that can be reached. The elimination of the aspect-dependent part of the measured signals is presented in Chapter 4, in order to extract the part of the signals containing aspect-independent parameters (i.e. the frequency of resonance f_r and quality factor Q). We use the PCB technology in Chapter 5 and inkjet printing in Chapter 6. For these reasons, this book can be differentiated from the existing works in the literature (Perret *et al.* 2012).

The idea of using chipless tags for authentication of manufactured products may not be conceived as RFID labels that can be attached, removed and then reattached to a product. Chipless RFID authentication can be considered as numerous high-level tools (e.g. high-level security seal) that need to be permanently affixed to the product, and any forgery of the tag would change the backscattered response, resulting in the product being characterized as a counterfeit. The chipless tag can also be inserted (or hidden) in the corrugated layers or the pulp of the product packing, or directly printed on a security document. To use chipless RFID tags for authentication, at least two measurements (one for the database and the other for the comparison at the time of authentication) are required. This necessity results in a high-level authentication method where authentication can only be done when the same tag is present in an intact manner. For such a system, repeatability of signals for a single tag is very important. For this reason, comparison between the multiple measurements of the same tag should always theoretically produce a similarity level of 1. On the other hand, comparison between the multiple measurements of two tags (different due to natural randomness) should always theoretically produce a similarity level of 0.

2.3. Conclusion

In this chapter, numerous authentication techniques from the literature were discussed. We mentioned that our proposed chipless authentication is a highly secure application as a result of natural process variations.

3

Methodology and Proof of Concept

3.1. Introduction

In this book, the proposed idea is that natural randomness can produce unique EM signatures from chipless RFID tags that can possibly be used for authentication. The concept of chipless RFID is extended to authentication where each tag must present a unique signature that can never be reproduced even if someone tries to copy the tag. The proposed technique can be considered as the first step towards chipless authentication of manufactured products. Initially, the proposed idea is proved by purposely applying the dimensional variations along the resonators that mimic natural randomness ΔP (i.e. inherent in the fabrication process) along the dimensional parameters of resonators in order to find the minimum variation needed. For this purpose, two resemblance metrics have been introduced and the performances of these metrics to discriminate the classical chipless tags have been discussed. This chapter is organized as follows:

– section 3.2 discusses some research work from the literature related to the natural process variations in the PCB technology and the inkjet printing technology;

– sections 3.3 and 3.4 show a procedure of authentication and a procedure of statistical analysis of the proposed approach, respectively, to present a methodology to characterize the chipless RFID tags for the application of authentication;

– sections 3.5 and 3.6 present a proof of concept by purposely applying the dimensional variations along the resonators for the PCB technology and the inkjet printing technology, respectively;

– section 3.7 concludes this chapter.

3.2. Randomness inherent in the realization process

For the PCB technology using a chemical etching process, non-systematic under- and over-etching are among the fabrication process anomalies (Khandpur 2006, Chap. 9; Pecht and Ganesan 2006). These non-systematic under- and over-etching can occur up to the thickness of microstrip metal trace t as provided by the design tolerance from a PCB fabrication company (Printed s.r.o.[1] in our case). It means that if $t = 35$ µm, then the natural dimensional randomness ΔP would also be in the order of 35 µm.

For the inkjet printing technology, the print quality is vitally dependent on the chosen resolution. Apart from this, numerous factors are also involved that determine the print quality, such as the design size and shape, ink density and type, substrate texture, drop position accuracy and type of image (Hutchings and Martin 2012, Chap. 2). For instance, a straight line can also have random variations along its edge (Duineveld 2003). As another example, the thickness of the inkjet-printed layer is in the order of a few micrometers (Nur *et al.* 2002; Redinger *et al.* 2004), as compared to 17.5 µm or 35 µm for conventional PCB microstrip trace. As another example, in Jung *et al.* (2007) and Salmerón *et al.* (2014a, 2014b), a layout, as shown in Figure 3.1, has been presented to explain a change in the width of inkjet-printed layer, where w is the width provided in the digital design and w' is the output printed width. It can be observed that the ink drops induced a width w' that is wider than the desired width w.

In the literature, numerous studies on the effect of process variations for RF circuits have been reported. In Batchelor *et al.* (2009), a frequency-selective surface (FSS) has been realized using inkjet printing. The printed dipoles show notable random dimensional variations in their print quality, where edge acuity has been found to be nonlinear

1 http://www.printed.cz/eng.

and also the thickness of the edges has been found as changing from 2.5 μm to 2 μm from edges to the middle of the trace, respectively. In Turki *et al.* (2013), detrimental effects in mass printing of FSS screens due to misprinting of square loop elements caused by the inkjet printing anomalies have been studied. It is found that in the absence of 20% square loop elements, the level of the transmission zero is reduced up to 16 dB and the frequency of resonance f_r exhibit a variation ranging from 450 MHz to 500 MHz.

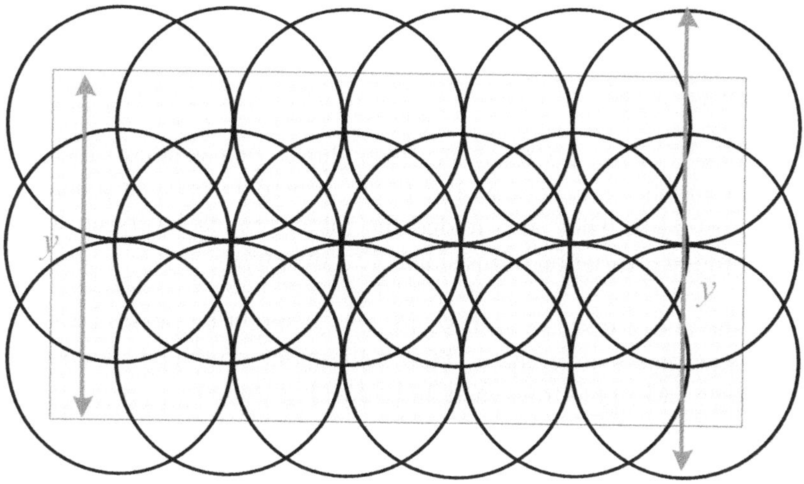

Figure 3.1. *A layout to explain a change in the width of the inkjet-printed layer from digital design to realization. For a color version of this figure, see www.iste.co.uk/ali/RFID.zip*

In Shao *et al.* (2009), process dependence of an inkjet-printed folded dipole antenna has been presented by studying the random variations of the width of the printed layer and the number of overprinting procedures. With a variation of the width of the printed layer from −20% to +20%, the gain increases from 1.73 dB to 1.77 dB and the read range increases from 0.99 to 1.01. On the other hand, the overprinting procedure does not contribute significantly to the performance of the antenna (e.g. the gain and radiation efficiency).

In Merilampi *et al.* (2010), the effect of the thickness of the inkjet-printed layer on the performance of printed quarter wave dipole has been investigated for the application of ultra-high frequency (UHF) RFID tag. It was found that the performance of the RFID antenna was deteriorated due to the reduction of the thickness of the inkjet-printed layer. Likewise, in Pynttari *et al.* (2012), the effect of the thickness of the inkjet-printed layer on the performance of coplanar waveguide (CPW) lines and two types of antennas has been studied. The attenuation loss of transmission lines and the total efficiency of antennas are improved by printing additional layers on the particular high-current-density areas. Furthermore, the efficiency of two-layer antenna is achieved with a partial second layer printed merely on high-current-density areas.

In Borghetti *et al.* (2016), it has been shown that conductivity linearly increases with the thickness of the inkjet-printed layer with a maximum variability of 13%. They also folded the inkjet-printed layers up to 1,000 times and found a variation of resistance less than 6%.

From the above discussion, it seems that it might be possible to exploit process randomness inherent in PCB and inkjet printing technologies to create unique EM signal from each individual realization.

3.3. Authentication procedure

This book proposes highly secure chipless authentication methods (see level 3 in Figure 2.2). In our proposed methods, a database of authenticity is made before deploying the authentication. This database of authenticity is the main reason for the high security level of the proposed method. Figure 3.2 presents the basic principle of the chipless authentication process. The authentication process based on chipless RFID tags is divided into two stages:

– measurement of the chipless RFID tags using a chipless reader and construction of a database (i.e. a pre-stage);

– comparison of a given unknown test tag with the database (i.e. a post-stage).

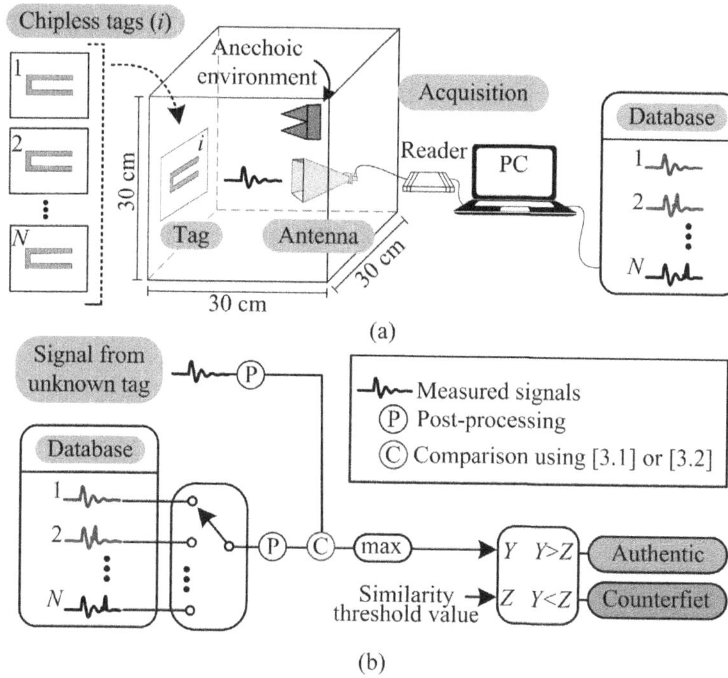

Figure 3.2. *Principle of the authentication procedure. (a) Pre-stage: formation of the database and (b) post-stage: comparison of a given unknown tag with the database. For a color version of this figure, see www.iste.co.uk/ali/RFID.zip*

Figure 3.2(a) illustrates the process of construction of the database, where each chipless tag is measured individually using a chipless RFID reader. Then, all measurements are stored in a database. For the measurements, a cubical box with a controlled anechoic environment is required. The size of the box can be, for example, 30 × 30 × 30 cm^3. By using such a small measurement box, we can ensure the consistency of the measurement environment, the distance between antennas if the setup is in bistatic configuration and the distance between the tag and antenna from the stage of construction of the database up to the deployment of authentication. Figure 3.2(b) illustrates the authentication procedure for a given unknown chipless RFID tag. First, all signals from the database as well as the signal from a given test tag are post-processed prior to the comparison. The post-processing consists of background normalization (i.e. subtraction of the measurement taken in the absence of chipless tag from the measurement taken in the presence of chipless tag) and time windowing (i.e. truncation of the signal in the time domain (TD) by zero padding to discard the structure made and to preserve the antenna mode). Further details of the post-processing will be presented in the discussion of measured signals in section 3.5.1. Then, the post-processed signal from the unknown test tag is compared to all post-processed signals from the database for finding a possible unique match. For the proposed comparison, the signals can be analyzed in the frequency domain (FD) using cosine similarity (CS) or in the TD using the maximum value of correlation coefficient (CC_{max}), as presented in Ali *et al.* (2018, 2019, 2020). The mathematical expression of CS is:

$$CS = \left| \frac{\Sigma(u_i \cdot v_i^*)}{\sqrt{\Sigma(u_i)^2}\sqrt{\Sigma(v_i)^2}} \right| \quad [3.1]$$

where $u = [u_1, u_2, \ldots, u_n]$ and $v = [v_1, v_2, \ldots, v_n]$ are two EM signals to be compared, exhibiting complex and discrete spectrums. The operator * represents the complex conjugate.

The mathematical expression of CC_{max} is:

$$CC_{max} = \max\left[\frac{\Sigma_i(n_i - \mu_n)(m_i - \mu_m)}{\sqrt{\Sigma_i(n_i - \mu_n)^2}\sqrt{\Sigma_i(m_i - \mu_m)^2}}\right] \quad [3.2]$$

where $n = [n_1, n_2, \ldots, n_n]$ and $m = [m_1, m_2, \ldots, m_n]$ are two time series EM signals to be compared, and μ_n and μ_m denote the arithmetic means of the two time series EM signals, respectively.

After the comparison of the signal from an unknown tag with the database, the maximum value of the resultant vector can be compared with a fixed similarity threshold value to check the authenticity of the tag. This similarity threshold is calculated from the statistical analysis that will be discussed in section 3.4. This similarity threshold should be chosen at the minimum probability of error, as it is a critical parameter that defines the performance of our approach.

A comparison between both similarity measures (CS and CC_{max}) is discussed in section 3.5.2, where we have applied these metrics to the measured signals.

From our viewpoint, a database for authentication is a primary requirement for an authentication technique to be a highly secure application. Therefore, an authentication approach without such a database-based comparison might not be considered as an application of level 3 of security.

3.4. Statistical analysis

In this book, the performances of the proposed methods are evaluated by using statistical analysis. Figure 3.3 presents the concept of the statistical analysis of the proposed approaches. The signals' comparison terms used in this book are the intra-tag, inter-tag and inter-realization. The intra-tag is referred to as the comparison among the repetitive measurements of the same chipless tag, and the inter-tag is referred to as the comparison among the different chipless tags. It is important to note that these different chipless tags are different in terms of realization from the same digital mask. Also, these tags can be on a same PCB. Inter-realization is referred to as the comparison among the different PCB realizations. These conventions are explained graphically in Figure 3.3(a). Apart from these, two other signals' comparison terms (intra-group and inter-group) are also used in this chapter. These conventions are explained in section 3.6.

From the comparison of the signals, the similarity distributions from all possible combinations among the signals of chipless tags are constructed. These distributions can be of any form or shape, as the comparison data is based on a random source. Ideally, the central tendency (e.g. mean) of intra-tag distribution should be towards a value of 1, as intra-tag distribution is based on the comparison among repetitive signals of each individual tag from the population. On the other hand, the ideal central tendency (e.g. mean) of inter-tag or inter-realization distribution should be towards a value of 0, as these distributions are based on the comparison among signals of different realizations of the tags from the population or different PCBs, respectively. For the sake of simplicity, intra-tag and inter-tag distributions are depicted only for realization 1 in Figure 3.3(a). However, of course, intra-tag and inter-tag distributions can also be constructed for realization 2.

Figure 3.3(b) presents the procedure for the calculation of error rates. With specifically chosen distributions (e.g. intra-tag and inter-tag or intra-tag and inter-realization) at an instant, the probability of false negative (PFN) and the probability of false positive (PFP) are computed. To make an example, we have chosen the intra- and inter-tag distribution from Figure 3.3(a) to Figure 3.3(b).

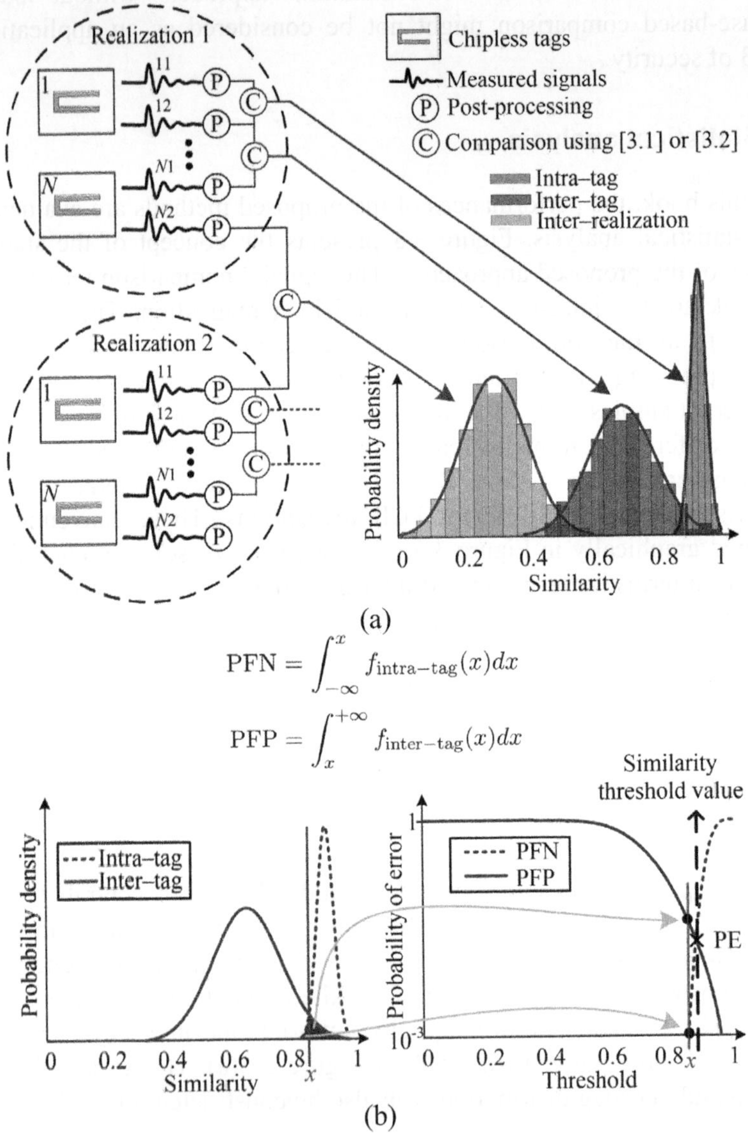

Figure 3.3. *The concept of statistical analysis of the proposed approach. (a) The procedure for the construction of the similarity distribution and also the description of signals' comparison conventions. (b) The procedure for the calculation of error rates. For a color version of this figure, see www.iste.co.uk/ali/RFID.zip*

By taking the intra-tag distribution, the PFN is calculated using the following expression:

$$\text{PFN} = \int_{-\infty}^{x} f_{intra-tag}(x)dx \quad [3.3]$$

where x corresponds to the threshold. We have presented an example of the calculation of the PFN from the intra-tag distribution with the threshold value x at an assumed position in Figure 3.3(b) (see the gray line from the intra-tag distribution to the PFN curve (dotted line)). In practice, the threshold value x is taken from 0 to 1 with 10,000 points to calculate the whole PFN curve (dotted line).

By taking the inter-tag or inter-realization distribution, the PFP is calculated using the following expression:

$$\text{PFP} = \int_{x}^{+\infty} f_{inter-tag\ or\ inter-realization}(x)dx \quad [3.4]$$

where x corresponds to the threshold. We have presented an example of the calculation of the PFP from the inter-tag distribution with the threshold value x at an assumed position in Figure 3.3(b) (see the gray line from the inter-tag distribution to the PFP curve (solid line)). In practice, the threshold value x is taken from 0 to 1 with 10,000 points to calculate the whole PFP curve (solid line).

Finally, the probability of error (PE) and the similarity threshold value (i.e. used above in section 3.3) are computed from the intersection point of the PFN and PFP.

3.5. Chipless tag discrimination using PCB tags

The idea is to change the dimensional parameters that are not directly linked to the frequency of resonance f_r in order to be able to estimate the classical variations in dimensions that can be measured with a common free space reflectometry approach in an anechoic environment. It is important to note that the design of the employed classical chipless tags is without any optimization on the shape for the issue of authentication. Chipless authentication will be discussed in Chapters 5 and 6 with special chipless RFID tags that have improved design for the authentication purpose.

3.5.1. *Chipless tag design and purposely applied dimensional variations*

The C-folded dipole has been extensively used in the design and development of chipless RFID tags (Perret 2014, Chap. 5). The response of such a tag is mainly driven by two key dimensional parameters (see Figure 3.4): the spacing between the two arms g and the length L′. The frequency of resonance f_r can then be written as:

$$f_r = \frac{c}{4L'\sqrt{\varepsilon_{eff}}} \tag{3.5}$$

where L' is the total length, which is the sum of L (physical length of each arm) and ΔL (complementary length added due to the fringing fields), and ε_{eff} is the effective permittivity of the substrate for a coplanar stripline (Garg *et al.* 2013, Chap. 7), which depends on the gap value g, the width w, the thickness t of the strips, the substrate thickness h and its permittivity ε_r. The calculation procedure of ε_{eff} for a coplanar stripline design is given in Appendix A. For the rest of this book, we use L' and L interchangeably for ease of discussion.

Each fabricated circuit being altered by ΔP exhibits a possible distinct signature. To evaluate the uniqueness of a tag, at least two measurements are needed: one for the database and the other for the examination. ΔP is an unknown random variable that critically depends on the technology chosen for circuit realization. For the conventional chemical etching process, this unknown ΔP is of the order of the metal thickness t. In this section, Rogers RO3003 substrate is used, having $\varepsilon_r = 3$, $h = 0.75$ mm and $t = 35$ µm. For a C-folded scatterer exhibiting dimensional parameters: $w = 2$ mm, $g = 0.8$ mm and $L' = 13.2$ mm, using Appendix A and [3.5], we calculate that $\varepsilon_{eff} = 1.2795$ and $f_r = 5.023$ GHz. As $t = 35$ µm, it is assumed that the maximum probable variation along the geometrical dimensions is 35 µm. If we impose an assumed maximum variation of 35 µm merely along the length, a significant shift of 13.4 MHz in the frequency of resonance f_r can be calculated using [3.5]. It is important to note that the frequency of resonance f_r of the scatterer is not directly linked to g and w (see [3.5] and Appendix A). The focus of this study is to examine the variations along these latter parameters (except L'), which have less effect on the frequency of resonance f_r. Thus, if we impose an assumed maximum variation on these two dimensions $w = 2 + 0.035$ mm and $g = 0.8–0.035$ mm, and keeping L' fixed, we deduce using Appendix A and [3.5] that $\varepsilon_{eff} = 1.2821$ and

f_r =5.018 GHz. This operation leads to a slight shift in the frequency of resonance f_r around an initial value of 5 MHz. Note that these variations would not merely affect f_r, but more generally, all the data points of the RF signature.

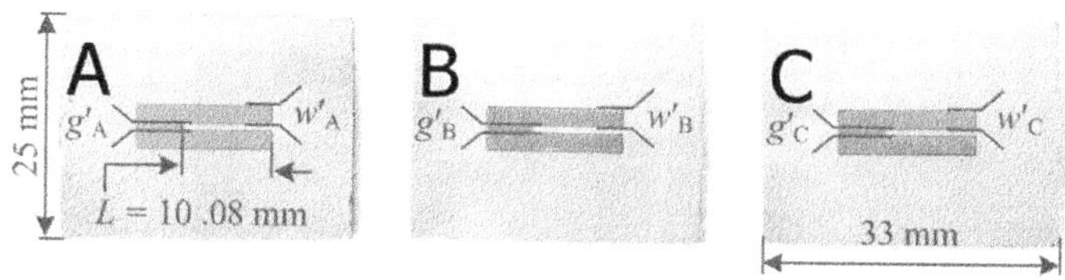

Figure 3.4. *Top view of fabricated PCB C-folded tags. For a color version of this figure, see www.iste.co.uk/ali/RFID.zip*

To experimentally validate the method, we fabricated three PCB chipless tags, as shown in Figure 3.4. Tag A exhibits the nominal theoretical geometrical dimensions, for example, $L = 10.08$ mm, $w = 2$ mm and $g = 0.8$ mm, designed for a resonance frequency of 5 GHz. In tag B, the spacing between the two arms g is decreased and the width of the trace w is increased by a purposely applied theoretical variation of 17.5 µm, whereas these theoretical variations are double for tag C. This chapter is based on the purposely applied variations, so the unknown ΔP occurring during the fabrication process is considered as a part of the applied variations. The purpose behind such applied variations is to prove the concept by introducing a proper analytical treatment that can quantify these very slight variations without ambiguity. Table 3.1 shows the dimensional parameters for all three fabricated tags, where the primed parameters w' and g' are the dimensional parameters, including the effect of the unknown ΔP. It is important to mention that each entry of ΔP in Table 3.1 shows a distinct and unknown process variation. Subsequently, we carefully measured the dimensions of the scatterers by an optical microscope to consider the unknown ΔP. These dimensions are measured at the two locations a and b on the C-folded scatterer, as shown in Figure 3.5.

Tags	Dimensional parameters (mm)	
	w'	g'
A	$w + \Delta P$	$g' + \Delta P$
B	$w + 0.0175 + \Delta P$	$g' - 0.0175 + \Delta P$
C	$w + 0.035 + \Delta P$	$g' - 0.035 + \Delta P$

Table 3.1. *Dimensional parameters of the three fabricated tags*

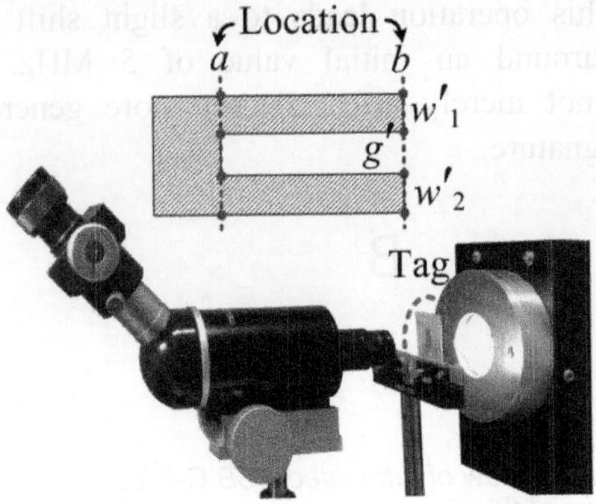

Figure 3.5. *Measurement of applied dimensional variations along parameters (w' and g') through an optical magnification system. For a color version of this figure, see www.iste.co.uk/ali/RFID.zip*

The average values of the measured dimensional variations among three tags in comparison to the theoretical applied variations are outlined in Table 3.2, where the contribution of ΔP can be observed. It can be observed that for both w' and g' parameters among three tags, the minimum measured average dimensional changes are $\Delta w'_{BC} = 7.89$ µm and $\Delta g'_{AB} = 4.6$ µm. All the other geometrical dimensions (e.g. L) of the C-folded dipole are the same for all the patterns, as well as the overall size of the tags: 33 × 25 mm^2.

Combinations among the tags	Measured average dimensional variations of the parameters (µm)		Theoretical applied variation of both parameters (µm)
	$\lvert\Delta w'\rvert$	$\lvert\Delta g'\rvert$	|Variation|
AB	23.68	4.6	17.5
BC	7.89	16.45	17.5
AC	31.58	21.05	35

Table 3.2. *Comparison of measured and theoretical applied dimensional variations*

Figure 3.6 shows the measurement setup in an anechoic environment with two antennas, keeping a spacing of $e = 2.7$ cm while the tag is placed at a distance of $r = 16.5$ cm from both antennas. The experimental results are measured using a VNA (Agilent 5222A) with a source power of –5 dBm.

All three tags are successively measured 15 times, where each measurement is done by removing and repositioning the tag at its position. Although measurements are carried out in a bistatic co-polarization configuration, we present only the results of reflection coefficient S_{11}, which seem to be more favorable for the practical case due to the requirement of fewer resources (one antenna). A large photograph of the measurement setup is also given in Appendix B.

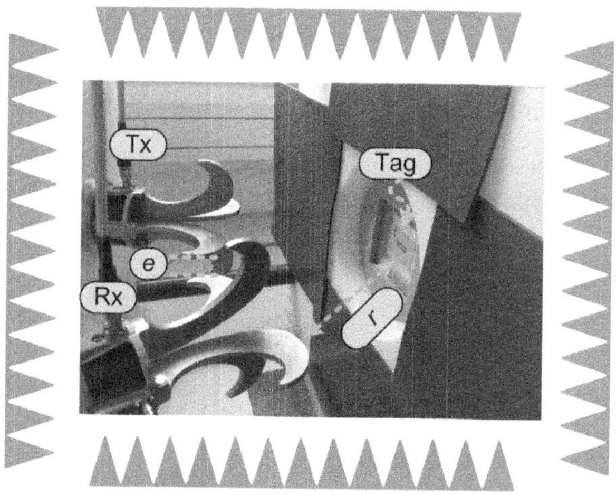

Figure 3.6. *Measurement setup in an anechoic environment. For a color version of this figure, see www.iste.co.uk/ali/RFID.zip*

For the post-processing, background normalization of S-parameters is carried out such that:

$$S^{tag} = S^{raw} - S^{back} \quad [3.6]$$

where S^{raw} and S^{back} are the measurements in the presence and absence of the tag, respectively. The FD and TD backscattered S_{11} of tag A are shown in Figure 3.7. The middle part of the TD signal is extracted by applying time windowing for a time duration of 40 ns (see Figure 3.7(b)). This time windowing is carried out by calculating the inverse fast Fourier transform (IFFT) of measured FD signal (non-windowed signals, solid black line), applying truncation by zero padding in TD and calculating the FFT of truncated TD signal (windowed signals, dotted gray line). The elimination of 15 ns of the early part of the TD signal is carried out to discard the direct reflections from the tag and its holder (the structural mode). These high reflections from the tag's holder are due to its solid plastic material. This solid holder is used to ensure the exact position in each measurement trial. It can be observed in Figure 3.7 that the peak apex related to the frequency of resonance f_r of the chipless tag is emphasized in the windowed signal because the structural mode and the early-time parasitic reflection are removed. A controlled measurement environment with a fixed distance between the tag and antennas (see Figure 3.2) ensures consistency of the time windowing. In this way, a selected time window to extract the antenna mode (and to discard the structural mode) would also remain the same from the phase of construction of the database up to the deployment of authentication.

For the rest of this section, we take into account the windowed signals for the similarity analysis, where both the FD and TD windowed signals exhibit 8,001 data points.

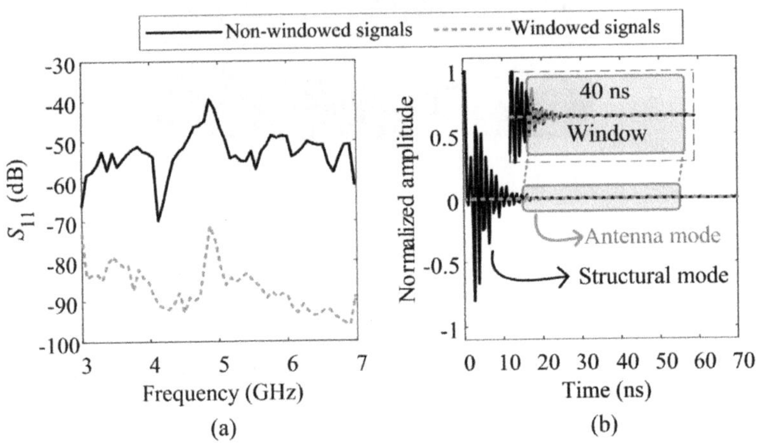

Figure 3.7. *Time windowing procedure for the reflection coefficient S_{11} of tag A. (a) Frequency domain responses and (b) time domain responses. For a color version of this figure, see www.iste.co.uk/ali/RFID.zip*

3.5.2. Chipless tag discrimination results and performance of the resemblance metrics

As for one tag, the total number of repetitive measurements is 15, the possible number of inter-tag combinations between two different tags is $15 \times 15 = 225$ and the possible number of intra-tag combinations to compare all repetitive measurements for each tag is $C_2^{15} = 105$. The level of similarity for all the combinations is calculated in two types of domains: in the FD using CS [3.1] and in the TD using CC_{max} [3.2].

The intra- and inter-tag CS distributions are presented in Figure 3.8(a), where Gaussian fitting is used to show the histogram distributions intuitively. The margin between the minimum intra-tag CS and the maximum inter-tag CS is approximately equal to 0.072. Figure 3.8(b) illustrates the intra- and inter-tag CC_{max} distributions calculated on all the backscattered signals of the three tags. Here, too, the Gaussian fitting is used to show the histogram distributions in an intuitive way. The margin between the maximum inter-tag CC_{max} and the minimum intra-tag CC_{max} is of about 0.138. The central tendency, for example, the mean value of the intra-tag distributions of both FD and TD analyses is close to unity (see Figure 3.8(a) and (b)), which validates the repeatability of each tag. A slight spread of the intra-tag distributions below unity is due to the uncertainties (systematic error and random error) from the measurement bench. On the other hand, the arrangement of the inter-tag similarity distributions of both FD and TD analyses is according to the measured average dimensional variations (see Table 3.2).

We calculated the PFN and PFP by using the fitted Gaussian probability density functions (see Figure 3.8(a) and (b)) for both FD [3.1] and TD [3.2] analyses. The error rates are presented in Figure 3.8(c). For the FD [3.1], we found a PE lower than 10^{-4}, when the closest distributions are chosen (i.e. the worst case): inter-tag B versus C distribution and intra-tag A distribution, while in the TD [3.2], the PE is lower than 10^{-10}, when the closest distributions inter-tag B versus C distribution and intra-tag B distribution are chosen (i.e. also a worst case).

Concerning the performance of similarity measures, the CC_{max} [3.2] outperforms the CS [3.1] as the margin is almost two-fold in TD analysis compared to FD analysis (see Figure 3.8). In general, both similarity measures (CS [3.1] and CC_{max} [3.2]) are robust. However, from a practical viewpoint where data acquisition is made in FD using a VNA, CS seems promising. Furthermore, as CS [3.1] is an absolute value normalized inner product between two complex vectors, it is less time-consuming compared to CC_{max} [3.2], where the correlation coefficient is calculated at each point by passing two real-valued vectors from each other and maximum value is taken. On the other hand, CC_{max} [3.2] being insensitive to the time delay in TD is also an efficient and robust alternative.

In Yang *et al.* (2016), a complex multi-resonant structure is used and the sum of the corresponding Euclidean distances among the multiple frequencies of resonance is used to distinguish the tags. However, in this chapter, a simple single-resonant structure is used and all the data points of the spectrum in both CS [3.1] and CC_{max} [3.2] are used to discriminate the tags. Hence, the performance of both CS [3.1] and CC_{max} [3.2] is good even if the contribution of the measurement bench is included (i.e. all the data points of the spectrum).

The concept discussed in this chapter would be applied to a larger number of tags with natural process variations for chipless RF authentication in Chapters 5 and 6.

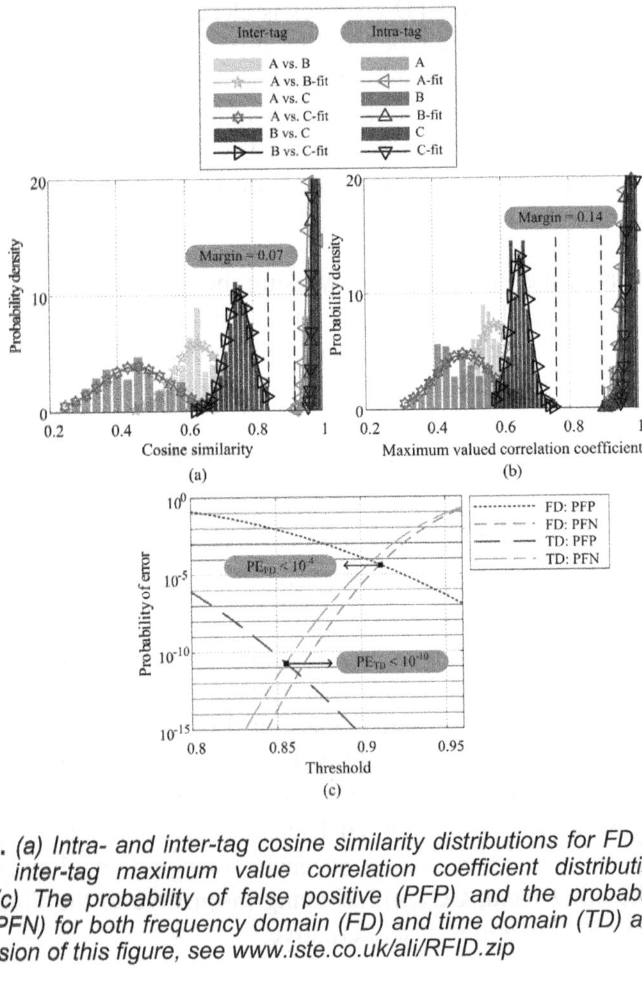

Figure 3.8. *(a) Intra- and inter-tag cosine similarity distributions for FD analysis. (b) Intra- and inter-tag maximum value correlation coefficient distributions for TD analysis. (c) The probability of false positive (PFP) and the probability of false negative (PFN) for both frequency domain (FD) and time domain (TD) analyses. For a color version of this figure, see www.iste.co.uk/ali/RFID.zip*

3.6. Chipless tag discrimination using inkjet-printed paper tags

The concept of applied dimensional variations, discussed in section 3.5, is also applied to test the capability of the tag discrimination using low-cost inkjet-printed paper-based tags.

3.6.1. *Chipless tag design and purposely applied dimensional variations*

Six identical tags for each identifier (i.e. A, B and C) are realized using inkjet printing. For the rest of this section, these six tags per identifier are referred to as group A, group B and group C. Figure 3.9 shows the top view of the first tag from each group. The geometrical dimensions for the C-folded scatterers for the groups A, B and C (see Figure 3.9) are exactly the same as the tags' geometrical dimensions for the C-folded scatterers for three tags, as shown in Figure 3.4 and Table 3.1. The tags are printed using an Epson EcoTank ET-2550 printer with JS-B25P silver conductive ink. The Epson premium glossy printable paper exhibiting thickness t = 270 μm is used as a substrate. We measured the relative permittivity $\varepsilon_r = 3.52$ and dielectric loss tangent tan δ = 0.12 of the substrate using the cavity resonator from Damaskos, INC. The measured tan δ for the Epson premium glossy printable paper is quite large compared to the Rogers RO3003 substrate. The overall size of the paper inkjet-printed tags is 48 × 48 mm^2. It is important to note that the design of the paper inkjet-printed tags includes the classical scatterer, that is, without any optimization on the design for the authentication.

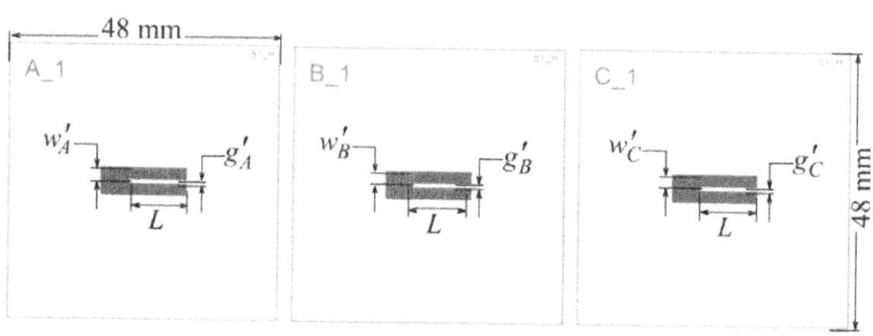

Figure 3.9. *Top view of paper inkjet-printed tags based on classical C-folded scatterer. w'_A = 2 mm + ΔP, g'_A = 0.8 mm + ΔP, w'_B = 2 mm + 17.5 μm + ΔP, g'_B = 0.8 mm − 17.5 μm + ΔP, w'_C = 2 mm + 35 μm + ΔP and g'_C = 0.8 mm − 35 μm + ΔP. For a color version of this figure, see www.iste.co.uk/ali/RFID.zip*

The reflection coefficient S_{11} of the paper inkjet-printed tags is measured in an anechoic environment using the measurement setup shown in Figure 3.6 with a distance from tag to antenna r = 3 cm. The reason behind reading the tags at such a short distance is that the paper substrate is highly lossy. Each tag from all three groups is measured five times successively, by removing and repositioning the tag at its position during each measurement trial to check the repeatability. Furthermore, a post-treatment of background normalization on measured signals is performed (i.e. subtraction of signal measured in the absence of the tag from the signal measured in the presence of the tag). The FD and TD backscattered S_{11} of the first tag of group A are shown in Figure 3.10. As the design of the scatterer is not optimized to be printed on the paper substrate, it can be observed that the peak apex corresponding to f_r is shifted from 5 GHz. For the time windowing, as explained in section 3.5.1, early signal up to 1.5 ns of TD signal is discarded and the subsequent signal of 10 ns is extracted to discard the structural mode and keep the antenna mode. For the rest of this section, we consider the windowed signals for the similarity analysis, where both the FD and the TD windowed signals exhibit 8,001 data points.

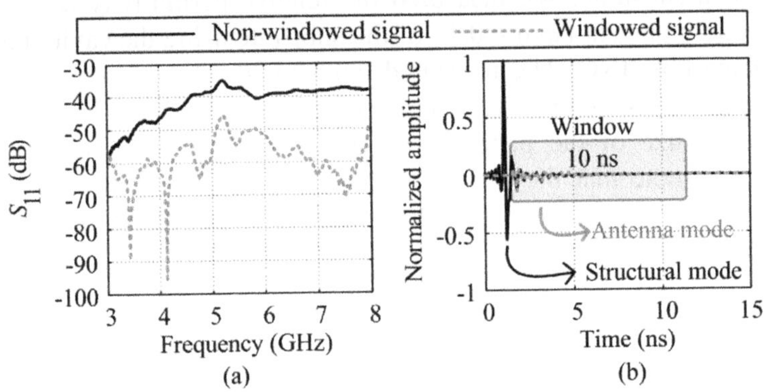

Figure 3.10. *Time windowing procedure for the reflection coefficient S_{11} of the first paper inkjet-printed tag of group A. (a) Frequency domain responses and (b) time domain responses. For a color version of this figure, see www.iste.co.uk/ali/RFID.zip*

3.6.2. Chipless tag discrimination results and performance of the resemblance metrics

Here, the signals are compared in terms of groups. Figure 3.11 depicts the signals' comparison conventions for the comparison among groups. The intra-group is referred to as the combinations to compare all measurements within one specific group (i.e. comparison among the repetitive measurements for each tag (intra-tag) and the comparison among all the individual tags within one group (inter-tag)). The inter-group is referred to as the combinations among the tags of two different groups of tags. Each group of inkjet-printed paper tags comprises six tags with each individual tag measured five times consecutively. So, each group has 30 measured signals. Then, the possible number of intra-group combinations is $C_2^{30} = 435$. The possible number of inter-group combinations is $30 \times 30 = 900$.

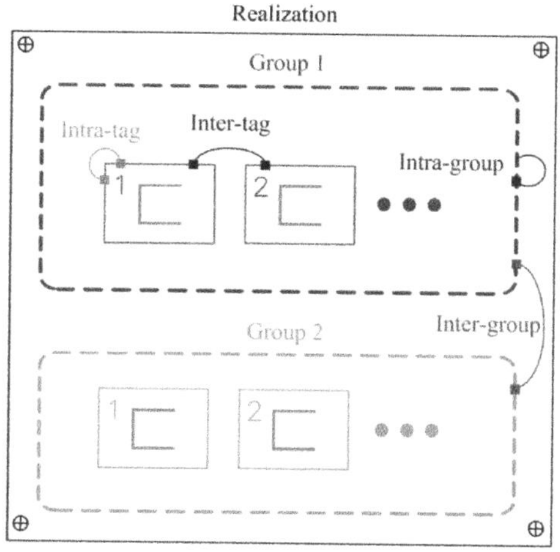

Figure 3.11. *Description of signals' comparison conventions for comparison among groups. For a color version of this figure, see www.iste.co.uk/ali/RFID.zip*

Figure 3.12 illustrates the similarity analysis for inkjet-printed paper tags, where FD analysis (Figure 3.12(a)) presenting the intra- and inter-group CS distributions are carried out using [3.1] and TD analysis (Figure 3.12(b)) presenting the intra- and inter-group CC_{max} distributions are carried out using [3.2]. The error rates calculated by using the similarity distributions are presented in Figure 3.12(c). The calculated overall system probabilities of error for the FD and TD analyses are equal to 32.45% and 32.64%, respectively.

Such large values of the probability of error are in contrast to the probability of error for classical PCB chipless tags (section 3.5). Although the geometrical dimensions of the scatterers among three groups (Figure 3.9) exhibit purposely variations in digital mask supplied to the printer, no discrimination among the inter-group distributions (Figure 3.12) is observed to yield a high probability of error. The reason behind this high probability of error can be explained by the microscopic image of the inkjet-printed paper tags. Figure 3.13 shows the microscope image of the upper arm of C-folded scatterer of the first tag of group A. The size of the ink particles varies from 50 μm to 120 μm that is larger than the maximum purposely applied variations 35 μm in the designs of the scatterers (i.e. from group A to group C). This implies that the resolution of the inkjet printer is lower than the variations existing in the digital mask of circuits of all three groups, and final printed prototypes might not exhibit any purposely applied dimensional variations. This indicates that the authentication with natural randomness in inkjet printing is very difficult without the development of an optimized design that is sensitive to inkjet printing.

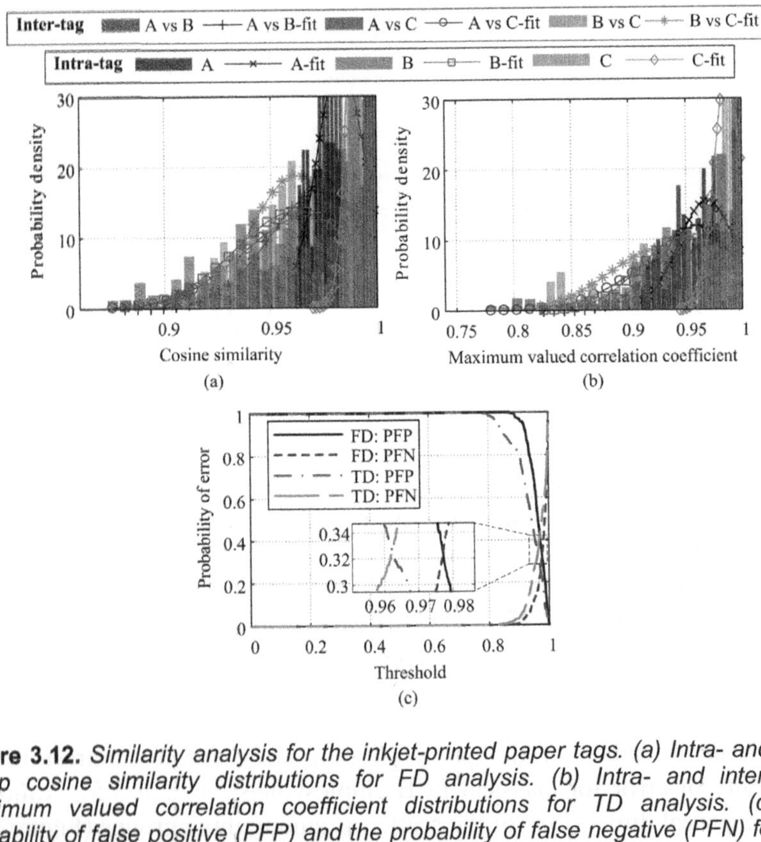

Figure 3.12. *Similarity analysis for the inkjet-printed paper tags. (a) Intra- and inter-group cosine similarity distributions for FD analysis. (b) Intra- and inter-group maximum valued correlation coefficient distributions for TD analysis. (c) The probability of false positive (PFP) and the probability of false negative (PFN) for both frequency domain (FD) and time domain (TD) analyses. For a color version of this figure, see www.iste.co.uk/ali/RFID.zip*

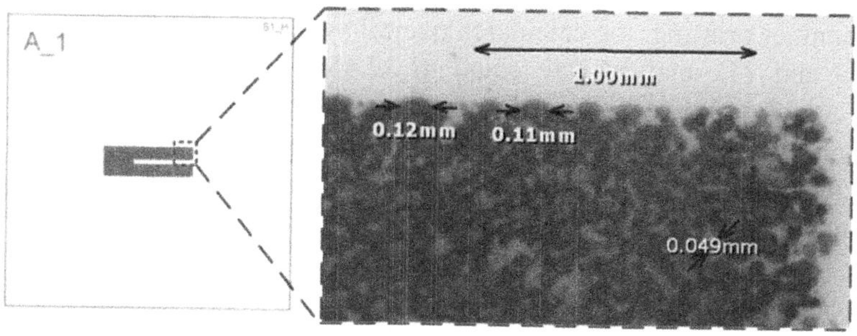

Figure 3.13. *Microscope image of the upper arm of C-folded scatterer of the first tag of group A. For a color version of this figure, see www.iste.co.uk/ali/RFID.zip*

3.7. Conclusion

A methodology to characterize the chipless RFID tag for authentication application was presented, where procedures to conduct authentication and to conduct statistical analyses are discussed.

The features of the chipless technology to be used for tag discrimination are demonstrated by using two technologies: PCB and inkjet printing. To validate the proposal, three chipless RFID tags were realized. Consecutively, from one tag to another, a variation (in the order of fabrication tolerance) was applied purposely to the geometrical dimensions exhibiting the lowest impact on the signal. Chipless tag discrimination based on the level of similarity was presented in both the frequency and time domains.

For PCB, the minimum measured average dimensional changes are $\Delta w'_{BC} = 7.89$ µm and $\Delta g'_{AB} = 4.6$ µm. In the FD analysis, we found a PE lower than 10^{-4}, when the closest distributions are chosen (i.e. the worst case): inter-tag B versus C distribution and intra-tag A distribution, while in the TD analysis, the PE is lower than 10^{-10}, when the closest distributions inter-tag B versus C distribution and intra-tag B distribution are chosen (i.e. also the worst case). Thus, these analyses show quite smaller values of probability of error even if the worst cases of intra- and inter-tag distributions were chosen. The PCB technology with natural random dimensional variations is very promising for use in authentication applications. Chipless authentication using PCB chipless tags is presented in Chapter 5.

For inkjet printing, the calculated overall system probabilities of error for the FD and TD analyses are equal to 32.45% and 32.64%, respectively. These large values of the probability of error are in contrast to the probability of error for classical PCB chipless tags. To use inkjet printing in authentication applications, natural random variations are not enough to generate significant unique signatures. For this reason, an optimization of the scatterer's design is strictly needed for inkjet printing to make this technology appropriate for authentication applications. Chipless authentication using inkjet-printed optimized chipless tags is presented in Chapter 6.

4

Extraction of Chipless Tag Key Parameters from Backscattered Signals

4.1. Introduction

Upon the emergence of the chipless RFID technology as a potential tool for item-level tagging (Perret 2014, Chap. 5), the requirement of robust detection techniques has appeared as one of the significant challenges for the practical implementation of the chipless RFID technology. Robust detection techniques are needed because the response of the chipless tag is generally very low compared to the backscattering response from surrounding unknown objects. Recently, the scientific community has focused on the research efforts to increase the coding capacity of chipless RFID tags (see Khan *et al.* 2016, Table III). In frequency-coded chipless RFID, the high coding capacity implies the increase in the number of peaks (associated with resonant scatterers) in the limited UWB. This congestion in the limited UWB would consequently narrow the allocated band to each peak that serves as a coding channel for each scatterer. With such narrow and closely positioned coding channels operating in a practical environment, the peaks may be shifted from their required position due to the superposition of reflections comings from the presence of unknown objects surrounding the chipless tag. These uncontrolled shifts in the positions of peak apexes may lead to a decoding error for high-density coded chipless tags. For this reason, the increase in the coding capacity may not be beneficial due to less robust detection techniques for the operation in a practical environment. Furthermore, it is much difficult to read the identification (ID) of a tag in chipless technology as compared to in ultra-high frequency (UHF) RFID technology, as the modulation scheme in time is not present in the chipless RFID technology. To date, most of the scientific work done in the chipless RFID field is based on the two measurements: tag measurement and empty measurement (i.e. in the absence of the tag). Background normalization is a post-procedure on the measurements that is based on subtraction of the empty measurement from the tag measurement. The background normalization may be very difficult in practical cases, where the background is varying rapidly (only systematic signal distortion coming, for example, from known and constant objects or for antenna coupling can be removed). A backscattered field from a chipless tag is a combined response including the antenna coupling, the tag's substrate, the tag's scatterer and the background (Ramos *et al.* 2016). As a first observation, for the above-mentioned practical measurement issues, we have no access to the tag's RCS but to the backscattered electric field, where, at best, part of the antenna coupling and background has been removed using the background normalization. The second observation is about how the tag ID is extracted from this signal. For the frequency domain chipless tag, the ID is directly linked to the known positions of the frequencies of resonances of the tag's scatterers (Ramos *et al.* 2016). It is obvious to detect these frequencies of resonances from the tag's RCS as it can be done in simulation. Indeed, these frequencies of resonances correspond to the peak apexes or the dips of the curve depending on the chosen reading polarization (co- or cross-polarization) or the presence or absence of tag's ground plane. Hence, in the literature, the same approach is applied to simulations as well as to measurement results to detect the tag ID. From the first observation, it is clear that this is not the correct way to proceed with measurement results where signals from different origins composed RF signature. Strictly speaking, in such a case, the peak apex does not correspond to the resonance frequency of scatterer. This means that the extracted value is not independent of the measurement setup. Therefore, the extracted position of the peak apex is susceptible to decoding error.

To employ chipless RFID tags for authentication, background normalization and a controlled environment (a box-like anechoic environment) are required for better repeatability of results. Even with background normalization and a controlled environment, the extraction of aspect-independent parameters would be needed:

– if the chipless tags employed for authentication are based on multi-scatterers;

– these aspect-independent parameters are the main information carrying the uniqueness of the tags coming from the natural process variations because the parameters are independent of the aspects of the measurement setup;

– fewer resources would be needed to save the aspect-independent parameters in the database of authenticity.

Hence, a robust technique for the extraction of aspect-independent parameters is also needed for chipless authentication.

We propose a simple approach to overcome the above-mentioned issues for both robust identification and authentication. The basic idea is linked to the fact that the backscattered field from the tag's scatterer originates late in time after the coupling and structural mode (i.e. direct optical reflections). This late time response is linked to CNRs or poles of the scatterers. Furthermore, singularity expansion method (SEM) (Baum *et al.* 1991) states that the CNRs are aspect-independent information (i.e. not dependent on the excitation or aspect). Such an aspect-independent nature of the pole(s) of a scatterer body (i.e. late time response) can be analyzed by SEM in the complex frequency plane (S-plane). The positions of poles in the S-plane are independent. Therefore, the late time response of chipless tag will be used in this chapter to extract the aspect-independent parameters for the identification (to obtain a better accuracy compatible with the reading of high-density coded tags) and chipless authentication.

In the literature, the MPM (Sarkar and Pereira 1995) is one of the methods to extract the CNRs of a scatterer to implement the SEM. For the characterization of a tag's ID in the chipless RFID technology using CNRs, MPM (Blischak and Manteghi 2011) and its variant STMPM (Rezaiesarlak and Manteghi 2013) have already been discussed. In Rezaiesarlak and Manteghi (2014), the STMPM is applied to extract the high-dense tag code. In Rezaiesarlak and Manteghi (2015a), the early-time and late-time modes of the transient response from multi-scatterer targets have been distinguished using the STMPM. On the other hand, a scalar method has been presented in Kracek (2019). This method is based on bistatic co-polarization measurement configuration. It will be shown later in this chapter that, compared to the proposed spectrogram method, the MPM (or STMPM) is not very simple to implement and is a time-consuming approach, which is not so compatible with real-time reading.

Recently, Ramos *et al.* (2016) have addressed the tag detection for the chipless RFID technology in different environments using a technique based on STFT. With this technique, the tag ID is extracted without background normalization (i.e. single measurement) by using an averaged late time signal. It has been shown that the technique is efficient even without background normalization operating in a realistic outdoor environment. Here, we want to show that the method can be extended to extract aspect-independent parameters such as the frequency of resonance and damping factor. The inevitability of such an extension is because the peak apexes corresponding to the resonant scatterers in the frequency-coded chipless RFID are well-separated in the spectrum. Therefore, this property can be further exploited to extend STFT for the estimation of the damping factors associated with the resonant scatterers. However, these parameters extracted by the extended STFT averaging method (spectrogram method) reflect the same information as CNRs. For the first time, we prove that this method can be used to obtain CNRs.

We propose a novel method for the extraction of aspect-independent parameters (that are analogous to CNRs). The proposed method is based on STFT, which is referred to as a spectrogram method. This chapter is organized as follows:

– section 4.2 explains the employed chipless tags along with the measurement setup used inside and outside of the anechoic chamber;

– section 4.3 presents the introduction of the technique and its comparison to the MPM using a second-order scatterer tag;

– section 4.4 discusses the extraction of the aspect-independent parameters for multi-scatterer tags, where an improvement of multiple dedicated averaging windows limited in both time and frequency is adopted to enhance the performance of the spectrogram method;

– section 4.5 presents a comparison of computation time durations between the proposed spectrogram method and the MPM;

– section 4.6 concludes this chapter.

4.2. Chipless RFID tags and measurement setup

In this chapter, the employed tags are the depolarizing RF-encoding particle (REP) tags (the tag based on dual-L dipoles and the tag based on shorted dipoles oriented at 45°) that have been presented in Vena *et al.* (2013c). The layouts of these tags are depicted in the inset of Figure 4.1. The geometrical dimensions of these depolarizing tags can be found in Vena *et al.* (2013c, Tables I and II).

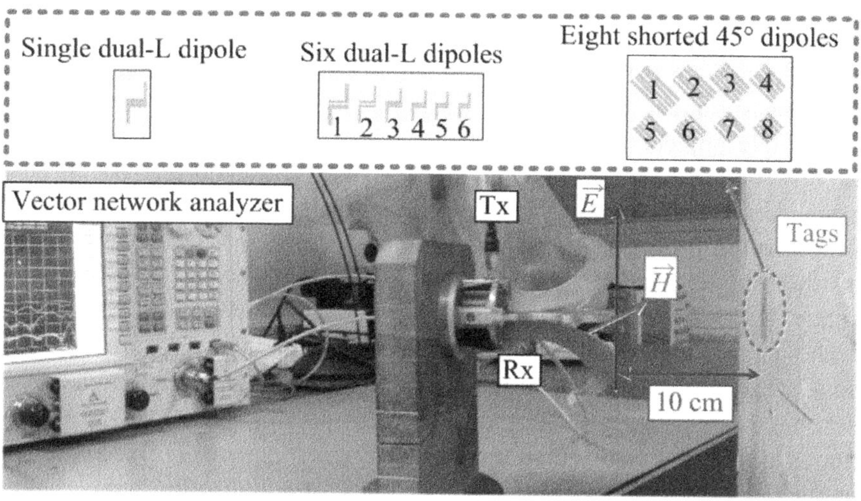

Figure 4.1. *Measurement setup used in an anechoic environment and in a realistic outdoor environment. The inset shows the layouts of the employed depolarizing chipless tags. The order of the scatterers is taken from the largest scatterer to the smallest scatterer. For a color version of this figure, see www.iste.co.uk/ali/RFID.zip*

The measurement setup used in an anechoic environment and in a realistic outdoor environment is shown in Figure 4.1. For the experimental results, an Agilent N5222A VNA with an output power of −5 dBm is used as a reader. The frequency sweep ranging from 3 to 8 GHz with 10,001 points is used. An open-boundary quad-ridge dual-polarization antenna (Satimo QH2000) is connected to VNA on ports 1 and 2, in vertical and horizontal polarizations, respectively. This antenna exhibits the gain approximately 12 dBi for the frequency sweep ranging from 3 GHz to 8 GHz. The isolation between ports of this antenna is greater than 30 dB. The distance between the tag and the antenna is 10 cm. The measured quantity is the transmission coefficient S_{21}.

In frequency-coded chipless RFID tags, the ID is normally coded with the presence or absence of the peak apex corresponding to the frequency of resonance f_{ri} of the ith scatterer. As explained in section 4.1, these peak apexes are aspect-dependent quantities that are affected by the operating environment. Therefore, in a realistic environment, these peak apexes do not occur precisely at the positions of f_{ri}. These uncontrolled shifts of peak apexes imply that the bandwidth of the coding channel associated with each scatterer must be enlarged to compensate these variations of the peak apexes. This action is eventually reducing the coding capacity. Otherwise, a decoding error may occur in the reading process. One solution of reducing the decoding error is to use background normalization (i.e. subtraction of the empty measurement from the tag measurement). Even so, in a practical case with mobile objects in the background, where the background is changing instantly or if the tag is attached to an unknown object, background normalization would not aid in the detection of tags. To show such changes in the reading process, for example, Figure 4.2 presents the measured $|S_{21}|$ for the single dual-L dipole tag without background normalization (Figure 4.2(a)) and with background normalization (Figure 4.2(b)) in an anechoic environment at two reader interrogation signal angles α_T: 15° and −15°. For measurements, the tag is attached to a foam support, which does not backscatter any part of the incident signal. The measurement setup is similar to a configuration where the reader antenna is fixed and only the tag can move. For the background normalization, the empty measurement (i.e. measurement in the absence of the tag) is measured at $\alpha_T = 0°$.

It can be observed from Figure 4.2(a) on both curves of the uncalibrated raw signals (i.e. without background normalization), the peak apexes linked to f_{ri} are not clearly visible due to the presence of the clutter (background information). As described in the analytical model introduced in section 4.5, we rather observe the presence of an attenuated peak followed by a dip which is characteristic of a superposition of at least two signals. If we consider the humps (that are followed by the dips) as peak apexes, the shift between them due to the change in α_T is approximately 15.5 MHz. The shift between the dips is 4 MHz. After removal of the clutter, the peak apexes corresponding to the f_{ri} are clearly visible in Figure 4.2(b) in the calibrated signals (i.e. with background normalization), exhibiting a shift of 24.5 MHz. Even so, it can be observed that due to the change in α_T (i.e. reading error due to disorientation), the dips associated with the peak apexes exhibit a shift of 80.1 MHz as α_T changes from 15° to –15°. Thus, the peak apexes of calibrated signals (Figure 4.2(b)) do not correspond precisely to f_{ri} of the scatterers, as these responses still contain the aspect-dependent information: coupling and structural mode (i.e. direct reflections). For this reason, post-processing techniques are required to extract the aspect-independent parameters for robust detection. Indeed, if such post-processing approach (i.e. background normalization and time windowing) is applied to the uncalibrated raw signals of Figure 4.2(a), a unique position of the peak apex equal to 4.16 GHz can be extracted, which thus can be considered to be the scatterer's frequency of resonance f_r = 4.16 GHz (ideal peak apex's position).

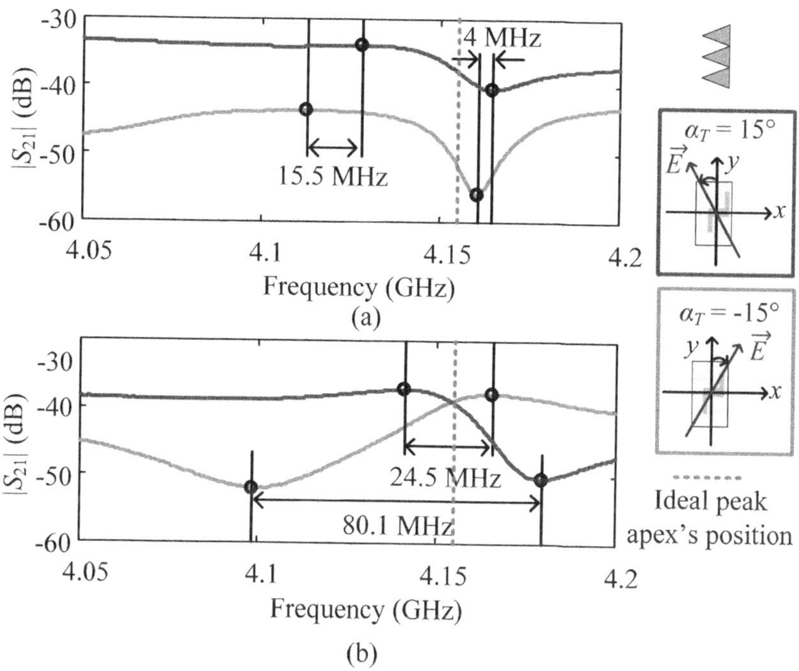

Figure 4.2. *Measured $|S_{21}|$ for the single dual-L dipole in an anechoic environment at two reader interrogation signal angles α_T: 15° and −15°. (a) Uncalibrated raw responses (without background normalization). (b) Calibrated responses (with background normalization). For a color version of this figure, see www.iste.co.uk/ali/RFID.zip*

In a tag reading process, the target tag is always known that can provide specific *a priori* information about the expected precise positions of the peak apexes corresponding to f_{ri}. In Figure 4.2, we have shown that the calibrated response (background-normalized) still contains aspect-dependent information. From our knowledge, in practice, the best way to discard the aspect-dependent information and then to extract the frequency of resonance f_r and the quality factor Q of the tag's resonant scatterers is as follows:

– for the measurement, to use an anechoic chamber and a VNA (with the parameters as mentioned above for Figure 4.1);

– for the post-treatment, to do a background normalization (subtraction of the empty measurement from the tag measurement, where between these two measurements no change in the reader configuration is done) and a time windowing (see section 3.5.1 or 4.3 for explanation) to suppress part of the early-time response where the signal-to-noise ratio is very low (i.e. the proportion of the tag response compared to, for example, the antenna coupling is low).

Such extracted parameters can be considered as the closest to those that we would have in simulation and very close to CNRs. Such a process has been used to characterize the tags that are used in this chapter for reference parameters. Thus, the positions of the peak apexes of signal from a tag processed by the above-discussed procedure can be used as reference *a priori* frequencies of resonances f_{riap} that correspond to f_{ri} to a larger degree. The reference *a priori* quality factors Qi_{ap} are extracted by using $Q_{iap} = f_{riap}/\text{BW}_{-3\text{ dB}}$, where $\text{BW}_{-3\text{ dB}}$ is the bandwidth at –3 dB. The reference *a priori* damping factors σ_{iap} can be calculated as found in Rezaiesarlak and Manteghi (2015b):

$$\sigma_{iap} = \frac{\omega_{riap}}{2Q_{iap}}, \qquad [4.1]$$

where ω_{riap} is the extracted *a priori* angular frequency of the *i*th scatterer.

The extracted *a priori* parameters associated with each scatterer of employed tags are outlined in Table 4.1, where the order of the scatterers is taken from the largest scatterer to the smallest scatterer (see the inset of Figure 4.1). The extracted information (see Table 4.1) will be used as the reference in the entire chapter.

Tags	*A priori* parameters	Quantities associated with all scatterers in a tag							
		1	2	3	4	5	6	7	8
Single dual-L dipole	f_{riap} (GHz)	4.16							
	Q_{iap}	109.34							
	σ_{iap} ($\times 10^8$)	1.18							
Six dual-L dipoles	f_{riap} (GHz)	4.33	4.70	5.22	5.77	6.34	6.88		
	Q_{iap}	98.39	99.65	104	95	82.67	88.52		
	σ_{iap} ($\times 10^8$)	1.38	1.48	1.58	1.91	2.41	2.44		
Eight shorted 45° dipoles	f_{riap} (GHz)	3.18	3.69	4.19	4.69	5.23	5.79	6.29	6.77
	Q_{iap}	82.42	97.1	87.66	61.93	61.41	43.92	43.74	38.32
	σ_{iap} ($\times 10^8$)	1.21	1.19	1.50	2.38	2.67	4.14	4.52	5.55

Table 4.1. *Extracted a priori parameters associated with each scatterer of the employed tags*

4.3. Extraction of aspect-independent parameters of a second-order scatterer

The backscattered field of a chipless tag can be modeled using a second-order bandpass filter model as discussed in Ali *et al.* (2018):

$$H(s) = X(s) + F(s) = \frac{\frac{\omega_0}{Q}\cdot s}{s^2+\frac{\omega_0}{Q}\cdot s+\omega_0^2} + F(s), \qquad [4.2]$$

where $s = j\omega$, $X(s)$ is a second-order bandpass filter model that is related to the resonating scatterer, $F(s)$ is a non-essential part of the response that may be the antenna coupling, the structural mode and the surrounding objects' response, and ω_0 and Q correspond to the angular frequency of resonance and quality factor. The surrounding objects can be resonant objects, but we assume that their corresponding frequencies of resonances f_{ri} are not within the frequency range of operation, and their corresponding quality factors Q_i are very small compared to Q of the chipless tag.

Using SEM, the resonating part $X(s)$ can be expanded to the poles and residues. The poles can be calculated from the denominator of $X(s)$, and the residues can be calculated by applying the partial fraction to $X(s)$:

$$H(s) = \frac{R_1}{s-s_1} + \frac{R_1^*}{s-s_1^*} + F(s), \quad [4.3]$$

where $s_1 = -\sigma_1 + j\omega_1$ (σ_1 is the damping factor and ω_1 is the angular frequency) is the CNR of a second-order pole, * denotes the complex conjugate, R_1 denotes the residue, σ_1, ω_1 and R_1 are $\sigma_1 = \frac{\omega_0}{2Q_1}$, $\omega_1 = \frac{\omega_0}{2}\sqrt{4 - \frac{1}{Q_1^2}}$ and $R_1 = \sigma_1(1 + j\frac{\sigma_1}{\omega_1})$, respectively, and Q_1 is the quality factor. For $Q_1 \geq 0.5$, the scatterer's frequency of resonance is equal to the natural frequency ω_0. For R_1, the angle θ_1 is $\theta_1 = tan^{-1}\frac{\sigma_1}{\omega_1}$. For [4.3], the inverse Laplace transform can be calculated as

$$h(t) = 2|R_1|u(t)e^{-\sigma_1 t}cos(\omega_1 t + \theta_1) + f(t), \quad [4.4]$$

where $h(t)$ is a TD response that is constituted of a damped sinusoid along with an uncharacterized response $f(t)$ that is related to antenna coupling, structural mode and surrounding objects' response, and $u(t)$ corresponds to the unit step function.

Similarly, the transfer function for a chipless tag based on multi-scatterers can be modeled as the following expression:

$$H(s) = \sum_{i=1}^{M} \frac{R_i}{s-s_i} + F(s), \quad [4.5]$$

where M is the number of scatterers and $s_i = \sigma_i + j\omega_i$ is CNR associated with ith mode. In a CNR, σ_i is the damping factor associated with the ith mode and ω_i is the angular frequency associated with the ith mode.

Then, TD backscattered field from a chipless tag based on multi-scatterers can be written as a sum of damped sinusoids including a summand for an uncharacterized response $f(t)$:

$$h(t) = \sum_{i=1}^{M} 2|R_i|u(t)e^{-\sigma_i t}cos(\omega_i t + \theta_i) + f(t). \quad [4.6]$$

It has been discussed in Ramos *et al.* (2016) that by applying time windowing to the backscattered TD signal from a tag, unwanted components (e.g. antenna coupling, structural mode and surrounding objects' response) can be discarded. Then, the antenna mode can be approximated as (Ramos *et al.* 2016):

$$h(t) \approx c_{vh,a}(t) \\ \approx Au(t-2\tau_R)e^{-\sigma(t-2\tau_R)}\cos(\omega_0(t-2\tau_R)), \quad [4.7]$$

where $c_{vh,a}(t)$ corresponds to the late time response of chipless tag, A determines the amplitude that depends on the coupling coefficient between emitted pulse and tag, as well as antennas effects, $u(t)$ corresponds to the unit step function that ensures causality and τ_R is the free-space propagation delay on the distance from antenna to tag. In the same way, the approximation of antenna mode for multi-scatterer-based tag can be written as

$$h(t) \approx \sum_{i=1}^{M} A_i u(t-2\tau_R)e^{-\sigma_i(t-2\tau_R)}cos(\omega_i(t-2\tau_R)). \quad [4.8]$$

Hence, [4.6] can also be approximated as [4.8]:

$$h(t) \approx \sum_{i=1}^{M} 2|R_i|e^{-\sigma_i t}\, cos(\omega_i t+\theta_i)\ T_{LB} \leq t \leq T_{LE}, \quad [4.9]$$

where T_{LB} and T_{LE} are, respectively, the beginning and ending times of the time window.

It has been discussed in Ramos *et al.* (2016) that the peak apexes corresponding to the f_{ri} can be extracted using the STFT averaging method. In this chapter, we are extending STFT averaging such that σ_i can also be extracted from the approximated response. We referred to this extended approach as the spectrogram method. It can be observed that the spectrogram method [4.8] and SEM [4.9] reflect the same information as CNRs (i.e. f_{ri} and σ_i). As SEM can be implemented using the MPM, the spectrogram method and the MPM provide the same information as CNRs. The accuracy of extracted parameters using the spectrogram method would also be compared to the accuracy of the extracted CNRs using the MPM later in this chapter.

Next, in this section, the modeling of a late time of a backscattered field by using the MPM is discussed and then the principle of our proposed spectrogram method is introduced by providing an analogy between the extracted parameters by using both the MPM and the spectrogram method. It is important to note that the uncalibrated raw signal (i.e. without background normalization) measured in an anechoic environment is used for the introduction of the proposed technique.

4.3.1. *Extraction with the matrix pencil method*

For the implementation of the total least-squares MPM, a data matrix [Y] is formed by using the TD signal h(t):

$$[\mathrm{Y}]=\begin{bmatrix} h(0) & h(1) & \cdots & h(L) \\ h(1) & h(2) & \cdots & h(L+1) \\ \vdots & \vdots & & \vdots \\ h(N-L-1) & h(N-L) & \cdots & h(N-1) \end{bmatrix}_{(N-L)\times(L+1)} \qquad [4.10]$$

where N is the total data samples in $h(t)$ and L is referred to the pencil parameter that is useful to eliminate the effect of noise from data. The optimum value of L should be chosen between $N/3$ and $N/2$. The MPM uses singular value decomposition (SVD) for the reduction of the dimensionality of the data [Y]. The singular values of the data [Y] are then used to define the desired number of poles M by comparing the ratio of each singular value δ to the largest singular value δ_{max} with a threshold value:

$$\frac{\delta}{\delta_{max}} = 10^{-p}, \qquad [4.11]$$

where p is the number of significant decimal digits.

The remaining steps of the algorithm of the MPM for calculating the complex poles and residues can be seen in Appendix C.

Figure 4.3 shows a flowchart of the extraction of physical poles by the MPM. The uncalibrated raw signal is used to extract CNRs of the single dual-L dipole tag (see the inset in Figure 4.1).

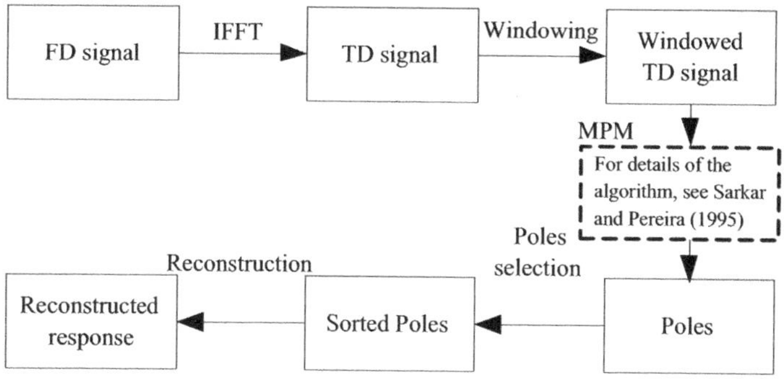

Figure 4.3. *Flowchart of the extraction of CNRs using the matrix pencil method*

Figure 4.4 shows the time windowing procedure to extract the sufficient late time response of the tag by applying a window of time duration 20 ns, where the beginning of the late time $T_{LB} = 2$ ns and the ending of the late time $T_{LE} = 22$ ns. The late time T_{LE} is estimated based on *a priori* extracted parameters (see Table 4.1). This time windowing is carried out by calculating inverse fast Fourier transform (IFFT) of measured uncalibrated raw FD signal (non-windowed signals, solid green line), applying truncation in TD and calculating FFT of truncated TD signal (windowed signals, dashed red line). The elimination of 2 ns (i.e. $T_{LB} = 2$ ns) of the early part of the TD signal is carried out to discard the direct reflections from the tag and its holder (the structural mode). Note that early-time elimination is mandatory to obtained accurate results. Windowed TD signal exhibiting 80 data points is used for the extraction of poles using the MPM with $p = 3$. Then, further sorting of the poles is carried out by applying a filtering bandwidth of 100 MHz $BW_{100\,MHz}$ in a range of $-50\text{ MHz} \leq f_{riap} \leq 50\text{ MHz}$. Figure 4.5(a) presents the CNR-based MPM reconstructed FD response in comparison with the supplied windowed FD response and the non-windowed FD response. The extracted CNRs using the MPM in comparison with *a priori* extracted parameters (like CNRs) are shown in Figure 4.5(b). MPM-based CNRs are shown in the complex conjugate form. For the rest of this chapter, only the positive CNRs would be shown for better visibility and better comparison.

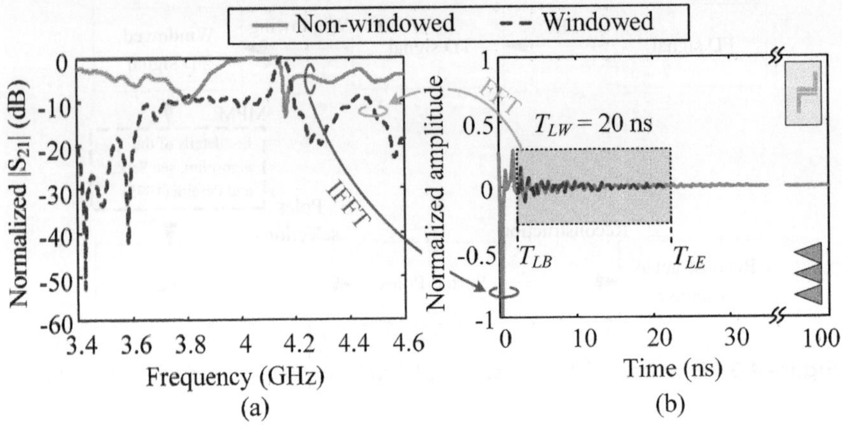

Figure 4.4. *Time windowing procedure for the single dual-L tag's uncalibrated raw response measured in an anechoic environment. (a) Frequency domain responses and (b) time domain responses. For a color version of this figure, see www.iste.co.uk/ali/RFID.zip*

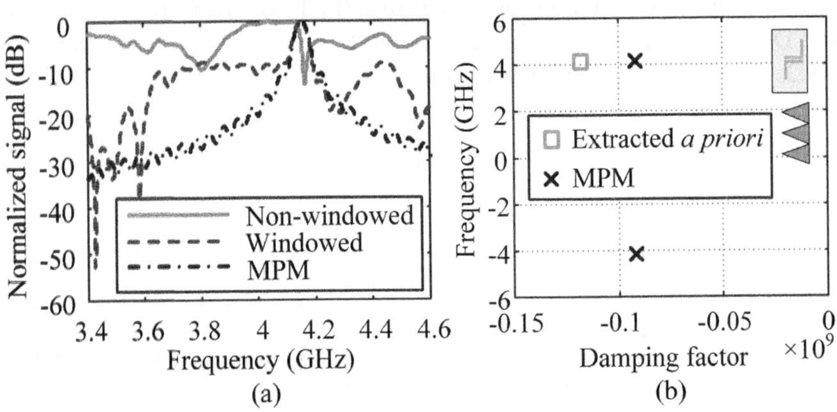

Figure 4.5. *Extraction of poles of the single dual-L dipole tag using the MPM measured in an anechoic environment with p = 3, T_{LW} = 20 ns and $BW_{100\ MHz}$ = –50 MHz ≤ f_{riap} ≤ 50 MHz and without the background normalization. (a) FD responses of reconstructed MPM response, non-windowed (uncalibrated raw) response and windowed (uncalibrated raw) response. (b) Extracted complex poles in comparison with extracted a priori complex poles (i.e. analogous to the CNRs; see Table 4.1). For a color version of this figure, see www.iste.co.uk/ali/RFID.zip*

4.3.2. Extraction with the spectrogram method

The basic definition of STFT of a given signal s(t) can be found in Allen (1982) and Gröchenig (2001, Chap. 4):

$$S(\omega,\tau) = \int_{-\infty}^{+\infty} s(t)w(\tau - t)\, e^{-j\omega t}dt, \qquad [4.12]$$

where $\omega = 2\pi f$ is the angular frequency, $w(t)$ is the weighting function or time window and τ is the time across sliding time widow $w(t)$.

Another interpretation of STFT discussed in Allen (1982) is that $S(\omega, \tau)$ can also be viewed as the output of a complex baseband filter bank, where ω is the center frequency of each bandpass filter and τ is the sampling time instant of the filter output. For the filter bank interpretation, the impulse response of the bandpass filter is

$$h_\omega(t) = w(t)e^{j\omega t}. \qquad [4.13]$$

Equation [4.12] can then be expressed as

$$\begin{aligned} S(\omega,\tau) &= w(t) * \left[e^{j\omega t}\, s(t)\right] \\ &= e^{-j\omega\tau}\int s(t)w(\tau - t)e^{-j\omega(t-\tau)}dt \\ &= e^{-j\omega\tau}\,[h_\omega(t) * s(t)]. \end{aligned} \qquad [4.14]$$

Thus, $S(\omega, \tau)$ can be perceived as a demodulated bandpass filtered response of $s(t)$.

The effect of the length of the window $w(t)$ to the trade-off of the resolution between time and frequency has been discussed in Ramos *et al.* (2016).

Figure 4.6 shows a flowchart of the extraction of physical poles by the spectrogram method. Figure 4.7(a) shows the calculated STFT for the single dual-L tag measured in an anechoic environment. The uncalibrated raw TD signal that is truncated at 100 ns (shown in Figure 4.4(b), solid green line) is supplied to the spectrogram method algorithm, where a Hamming window of 30 ns with 90% overlap is used to compute STFT.

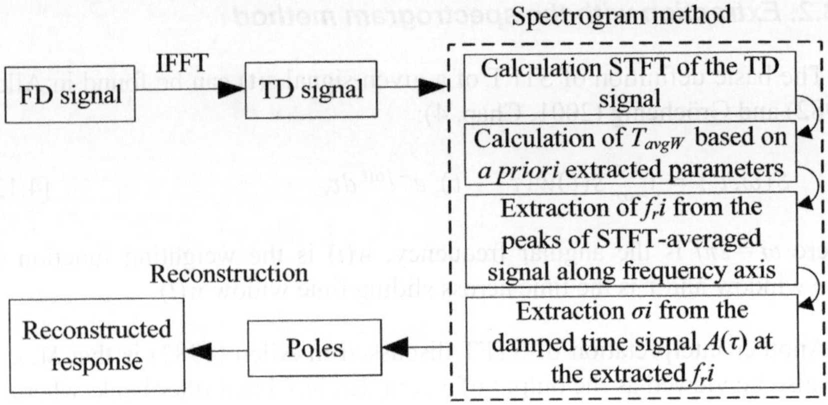

Figure 4.6. *Flowchart of the spectrogram method*

Figure 4.7(a) also shows a usable area (black dashed lines) for STFT averaging. To estimate this usable area for STFT averaging, the time width of STFT averaging window T_{avgW} can be calculated from the decaying envelope generated by using *a priori* extracted damping factors σ_{iap} (see Table 4.1):

$$a_{en} = e^{-\sigma_{iap}\, t_{en}}. \quad [4.15]$$

Then, [4.15] can be rearranged as

$$t_{en} = \frac{\ln(a_{en})}{-\sigma_{iap}}, \quad [4.16]$$

where σ_{iap} is *a priori* extracted damping factor and a_{en} is the amplitude of the decaying envelope. For intuition, we have shown the decaying envelope generated by [4.15] using *a priori* extracted damping factors σ_{iap} associated with the single dual-L dipole tag (see Table 4.1) in Figure 4.8. By fixing the amplitude of the decaying envelope to a threshold value a_{en} = 0.001 in [4.16], it would be certain that the mode has sufficiently decayed. So, the ending time of STFT averaging window can be taken as $T_{avgE} = t_{en}$ when a_{en} = 0.001. Similarly, T_{avgE} is calculated from [4.16], which is approximately 57.9 ns, as shown in Figure 4.7(a).

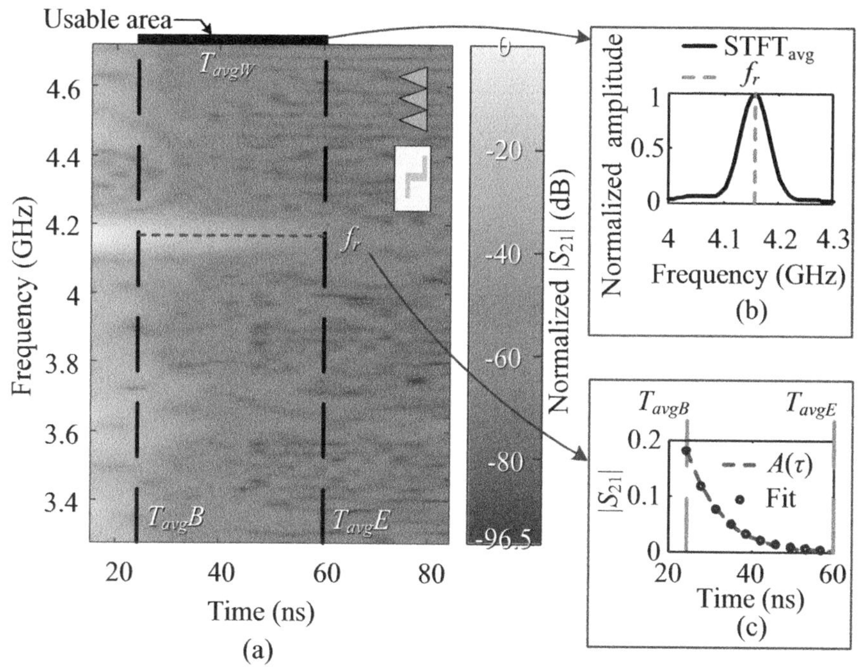

Figure 4.7. *Extraction of CNR by the spectrogram method for the single dual-L dipole tag using uncalibrated raw signal measured in an anechoic environment. (a) STFT calculated using a Hamming window of 30 ns with an overlap of 90%. (b) Extraction of f_r from the peak apex of $STFT_{avg}$ signal. (c) Extraction of σ from the damping time signal $A(\tau)$ selected at f_r inside T_{avgW}. For a color version of this figure, see www.iste.co.uk/ali/RFID.zip*

The beginning time T_{avgB} of STFT averaging window can be defined with respect to τ of $w(t)$ used in the calculation of STFT. This T_{avgB} can be defined as first-time instant of τ or after a few early-time instants of τ to avoid coupling (Ramos *et al.* 2016). T_{avgB} is fixed at 25 ns to avoid the coupling in the averaging of STFT. This choice of T_{avgB} = 25 ns is similar to the T_{LB} = 2 ns used to discard the structural mode in Figure 4.4. It is important to note that the difference between the extracted parameters using the spectrogram approach (Figure 4.7) and *a priori* extracted parameter (Table 4.1) is that the TD signal used for the extraction in the spectrogram approach is uncalibrated (i.e. without background normalization) and without the time windowing procedure.

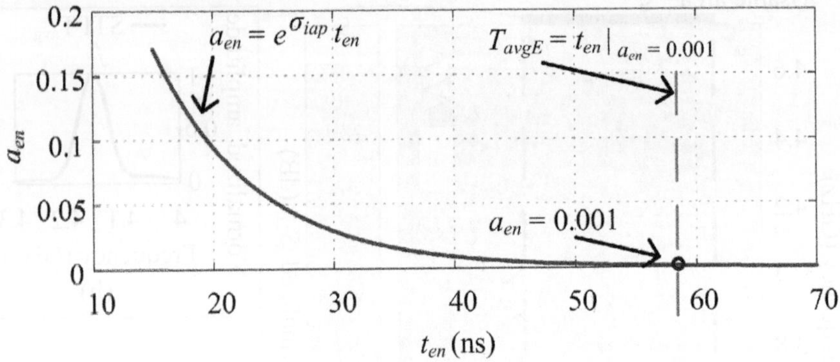

Figure 4.8. *Calculation of parameters of STFT averaging window from the decaying envelope generated by using the extracted a priori damping factors σ_{iap} at a fixed amplitude of the envelope threshold value a_{en} = 0.001. For a color version of this figure, see www.iste.co.uk/ali/RFID.zip*

Then, the column vectors of STFT in the range of T_{avgW} (usable area) are averaged, as shown in Figure 4.7(a). From the viewpoint of the flowchart of the proposed technique (Figure 4.6), f_{ri} related to the ith scatterer can be extracted from the peak apexes of the STFT averaged ($STFT_{avg}$) signal. In the same way, Figure 4.7(b) shows the extraction of f_r from the peak apex of the $STFT_{avg}$ signal for the single dual-L dipole tag.

Further, according to the flowchart of the proposed technique (Figure 4.6), for a multi-scatterer chipless tag, σ_i related to the ith scatterer can be calculated from the decaying time vector $A(\tau)_i$ residing inside T_{avgB} at the extracted f_{ri} (i.e. the position of the peak apex in the $STFT_{avg}$ signal). The calculation of σ_i is carried out by the least-squares method of solving an over-determined linear system of equations of the exponential model of $A(\tau) = A_o \exp(\sigma_i\, \tau)$ for amplitude A_o and damping factor σ_i. For the single dual-L dipole tag, the extraction of σ from the damping time signal selected at f_r inside T_{avgW} is presented in Figure 4.7(c).

Finally, a second-order bandpass filter model (a simple form of [4.2]) is used for the reconstruction, as discussed in Ali *et al.* (2018):

$$H(\omega) = \frac{\frac{2\sigma j\omega}{\omega_0^2}}{1+\frac{2\sigma j\omega}{\omega_0^2}+\left(\frac{j\omega}{\omega_0}\right)^2}, \qquad [4.17]$$

where ω_o and σ correspond to the angular frequency of resonance and the damping factor of a pole of second order, respectively.

In the rest of this chapter, for the proposed spectrogram approach, the position of peak apex in the STFT_{avg} signal and the frequency of resonance f_{ri} corresponding to each scatterer will be used interchangeably. The extracted f_{ri} and σ_i parameters by the proposed spectrogram method are analogous to the poles (CNRs) extracted by using the MPM.

Figure 4.9(a) shows a comparison of reconstructed FD responses from both the MPM and the spectrogram method along with their corresponding supplied uncalibrated raw responses (i.e. windowed FD response and non-windowed FD response). The uncalibrated raw windowed TD response and uncalibrated raw non-windowed TD response are supplied to the MPM and the spectrogram method, respectively, to extract the aspect-independent parameters for the single dual-L dipole tag measured in an anechoic environment. A zoom of Figure 4.9(a) is provided to show the matching between reconstructed responses from both employed methods. Figure 4.9(b) shows a comparison of extracted complex poles and *a priori* extracted complex pole. The difference of the MPM extracted parameters from *a priori* extracted parameter is $|\Delta f_r^{MP}| = 0.57$ MHz, $|\Delta\sigma^{MP}| = 0.02 \times 10^9$ and $|\Delta Q^{MP}| = 32.68$. The difference of the spectrogram extracted parameters from *a priori* extracted parameter is $|\Delta f_r^{SM}| = 1.54$ MHz, $|\Delta\sigma^{SM}| = 0.0006 \times 10^9$ and $|\Delta Q^{SM}| = 0.48$.

We can argue that the extracted $|\Delta Q^{MP}|$ is very large, such that the accuracy of the MPM is questioned. Strictly speaking, the MPM is dependent on T_{LB}. If we use an optimized value of $T_{LB} = 17$ ns instead of $T_{LB} = 2$ ns (i.e. as used above), the performance of the MPM is improved as $|\Delta Q^{MP}| = 2.75$. It is very important to mention that the application of the MPM is not direct, but it needs some skills such as choice of T_{LB}, T_{LE}, M and p. If care is not taken, numerous spurious poles can emerge, even after applying a filtering bandwidth of 100 MHz $\text{BW}_{100\text{ MHz}}$ in a range of $-50\text{ MHz} \leq f_{riap} \leq 50\text{ MHz}$.

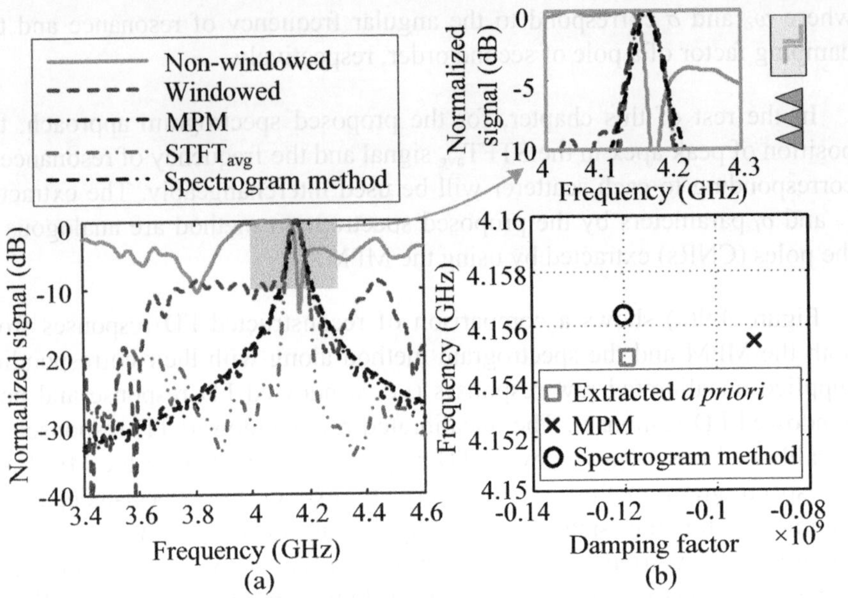

Figure 4.9. *Comparison of the MPM and the spectrogram method. (a) Reconstructed FD responses along with their corresponding supplied uncalibrated raw responses (i.e. windowed FD response and non-windowed FD response). The uncalibrated raw windowed TD response and uncalibrated raw non-windowed TD response are supplied to the MPM and the spectrogram method, respectively, where they are used to extract the aspect-independent parameters for the single dual-L dipole tag measured in an anechoic environment. (b) Extracted complex poles in comparison with a priori extracted complex pole. For a color version of this figure, see www.iste.co.uk/ali/RFID.zip*

It is realistic to imagine that we know the tag type to be detected and *a priori* information can optimize the processing of both the MPM and the proposed technique. However, it is important to note that the spectrogram technique does not become entirely dependent on *a priori* extracted parameters for the calculation of STFT averaging window parameters. These parameters can also be decided without the information of *a priori* parameters. We have tested that even if STFT averaging window parameters are taken as $T_{avgB} = 20$ ns and $T_{avgE} = 50$ ns, we found that the extracted f_{ri} and σ_i parameters are exactly the same as presented in Figure 4.9 for the spectrogram method.

4.4. Extraction of CNRs of the multi-scatterer-based tags

For extraction of aspect-independent parameters for the multi-scatterer-based tags (see Figure 4.1) using the spectrogram method, the algorithm follows the same steps as explained in Figure 4.6 (section 4.3.2). It is important to note that here, like in the previous section, the measured responses of the six dual-L dipoles tag supplied to the spectrogram method are uncalibrated raw TD signals.

Figure 4.10 shows the extraction of CNRs by the spectrogram method for the six dual-L dipoles tag measured in an anechoic environment. The calculated STFT of the uncalibrated raw TD signal is shown in Figure 4.10(a), where a Hamming window of 30 ns with 90% overlap is used to compute STFT. In this multi-scatterer tag detection, T_{avgW} is calculated by using *a priori* extracted parameters of the largest and the smaller scatterers, where the order of the scatterers is taken from the largest scatterer ($n = 1$) to the smallest scatterer ($n = 6$), as shown in Figure 4.10. Using [4.16] along with a threshold amplitude $a_{en} = 0.001$, the parameters of STFT averaging window are calculated as $T_{avgE} = t_{en}^{\sigma_{1ap}}$, that is, 49.93 ns and $T_{avgB} = t_{en}^{\sigma_{6ap}} - 6$ ns, that is, 22.30 ns. The reason behind the subtraction of 6 ns from $t_{en}^{\sigma_{6ap}}$ in the calculation of T_{avgB} is to provide a sufficient time to extract the damping factor because $t_{en}^{\sigma_{6ap}}$ is the time at which the envelope reaches a threshold amplitude $a_{en} = 0.001$. The above-discussed method is one way to calculate the optimized parameters of the STFT averaging window. Otherwise, these parameters can also be fixed to certain values. It will be shown later in this section, where we have used T_{avgB} with a fixed $T_{avgB} =$ 15 ns for the extraction of CNRs by the spectrogram method for the six dual-L dipoles tag using the uncalibrated raw signal measured in a realistic outdoor environment.

Figure 4.10(b) shows the extraction of f_{ri} from the peak apexes of the STFT_{avg} signal. The extraction of σ_i is carried out from the damping time $A(\tau)_i$ signal selected at each f_{ri} inside T_{avgW}. The extraction of σ_6 and σ_1, for example, is presented in Figure 4.10(c) and (d), respectively.

For the extraction of aspect-independent parameters for the multi-scatterer-based tags (see Figure 4.1) using the MPM, the algorithm follows the same steps as explained in Figure 4.3 (section 4.3.1). Time windowing is applied to the uncalibrated raw TD signal measured in an anechoic environment with $T_{LW} = 16$ ns, where an early part up to $T_{LB} = 2$ ns is discarded, as explained in Figure 4.4. The windowed TD signal exhibiting 160 data points is supplied to the MPM with $p = 3$ to extract CNRs. Then, further sorting of the poles is carried out, applying a filtering bandwidth of 100 MHz $BW_{100\ MHz}$ in a range of $-50\ MHz \le f_{riap} \le 50$ MHz.

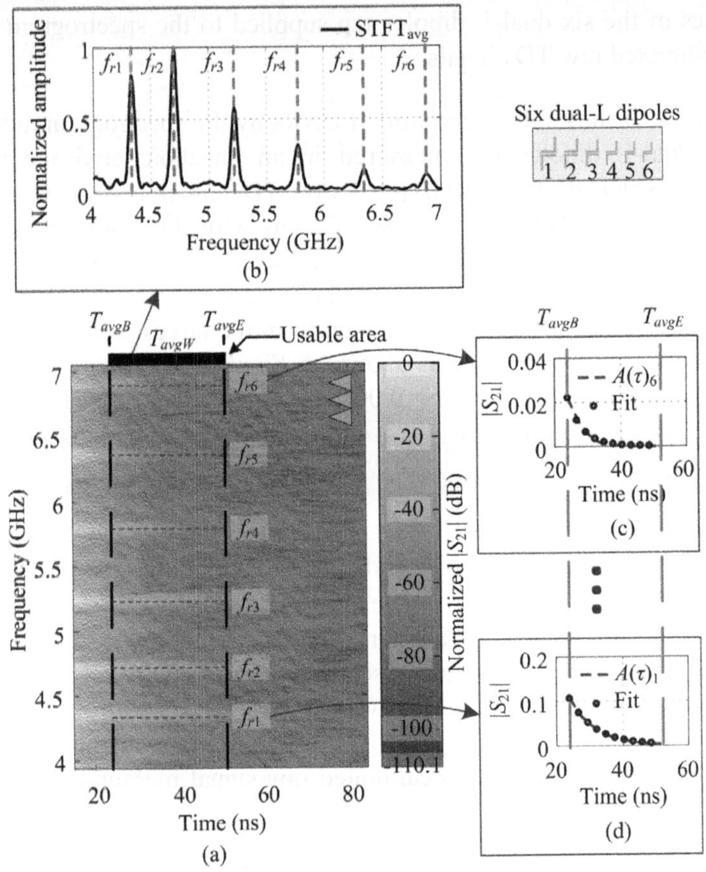

Figure 4.10. *Extraction of CNRs by spectrogram method for the six dual-L dipoles tag using the uncalibrated raw signal measured in an anechoic environment. (a) STFT calculated using a Hamming window of 30 ns with an overlap of 90%. (b) Extraction of f_{ri} from the peak apexes of $STFT_{avg}$ signal. (c) and (d) Extraction of σ_6 and σ_1 from the damping time signals $A(\tau)_6$ and $A(\tau)_1$ selected at f_{r6} and f_{r1} inside T_{avgW}, respectively. For a color version of this figure, see www.iste.co.uk/ali/RFID.zip*

Figure 4.11(a) shows a comparison of reconstructed FD responses from both the MPM and the spectrogram method along with their corresponding supplied uncalibrated raw responses (i.e. windowed FD response and non-windowed FD response). The uncalibrated raw windowed TD response and uncalibrated raw non-windowed TD response are supplied to the MPM and the spectrogram method, respectively, to extract the aspect-independent parameters for the six dual-L dipoles tag measured in an anechoic environment. Figure 4.11(b) shows a comparison of extracted complex poles and *a priori* extracted complex poles.

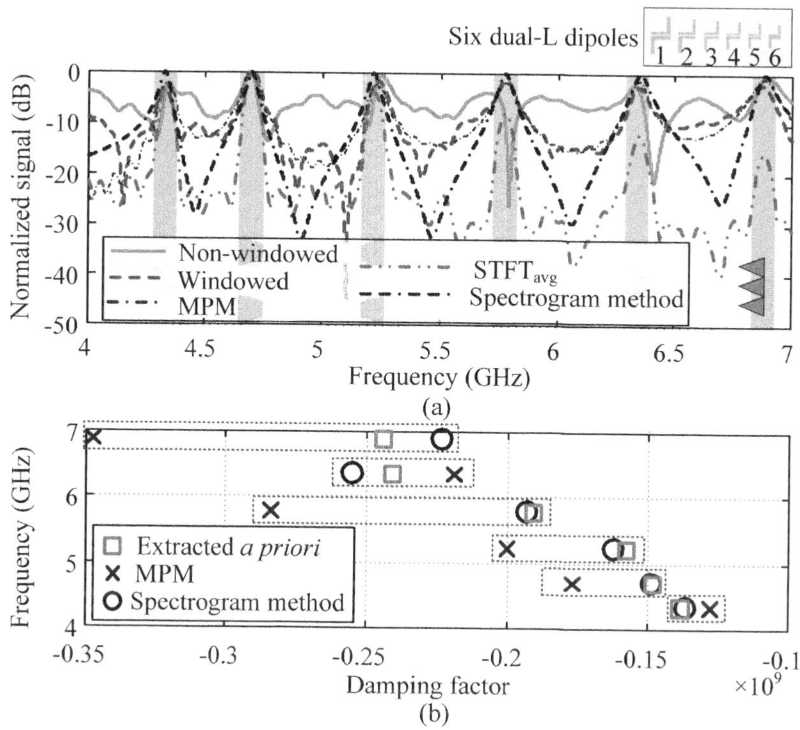

Figure 4.11. *Comparison of the MPM and the spectrogram method. (a) Reconstructed FD responses along with their corresponding supplied uncalibrated raw responses (i.e. windowed FD response and non-windowed FD response). The uncalibrated raw windowed TD response and uncalibrated raw non-windowed TD response are supplied to the MPM and the spectrogram method, respectively, where they are used to extract the aspect-independent parameters for the six dual-L dipoles tag measured in an anechoic environment. (b) Extracted complex poles in comparison with a priori extracted complex poles. Shaded areas are in the range of* -50 *MHz* $\leq f_{riap} \leq$ *50 MHz. For a color version of this figure, see www.iste.co.uk/ali/RFID.zip*

The absolute differences of extracted parameters using both the spectrogram method and the MPM from their corresponding *a priori* extracted parameters are outlined in Table 4.2. It can be observed from Figure 4.11 and Table 4.2 that the spectrogram method outperforms the MPM for the majority of scatterers.

Scatterer *i*	**Spectrogram method**			**Matrix pencil method**		
	$\|\Delta f_{ri}\|$ **(MHz)**	$\|\Delta\sigma_i\| \times 10^9$	$\|\Delta Q_i\|$	$\|\Delta f_{ri}\|$ **(MHz)**	$\|\Delta\sigma_i\| \times 10^9$	$\|\Delta Q_i\|$
1	≈ 0	0.002	1.48	3.90	0.01	8.13
2	≈ 0	≈ 0	0.23	4.27	0.028	16.25
3	≈ 0	0.005	2.85	9.6	0.042	21.92
4	≈ 0	0.002	0.98	3.78	0.093	31.04
5	9.76	0.013	4.36	3.67	0.022	8.37
6	≈ 0	0.02	8.29	2.74	0.103	26.3

Table 4.2. *Absolute differences of extracted parameters using the spectrogram method and the MPM from the corresponding* a priori *extracted parameter – anechoic environment*

Similar to the scheme presented in Figure 4.10, the extraction of CNRs by the spectrogram method for the six dual-L dipoles tag measured in a realistic outdoor environment is carried out, as shown in Figure 4.12. STFT of the uncalibrated raw TD signal is calculated, as shown in Figure 4.12(a), where T_{avgB} is fixed at 15 ns (i.e. beginning of the time τ across the sliding window), as in a noisy outdoor case, the signal-to-noise ratio (SNR) is low. Figure 4.12(b) shows the extraction of f_{ri} from the peak apexes of STFT_{avg} signal. The extraction of σ_i is carried out using the damping time signal $A(\tau)_i$ selected at each f_{ri} inside T_{avgW}. The extraction of σ_1, for example, is presented in Figure 4.12(c). In a noisy outdoor case, outlier peaks can also emerge in the detection process, as shown in Figure 4.12(b). Such outliers can be avoided by observing a monotonic decreasing trend of the TD signal at the selected peak. The peak apex, for example, shown in FD in Figure 4.12(b) is identified as an outlier, as it does not exhibit a monotonic decreasing trend in TD, as shown in Figure 4.12(d).

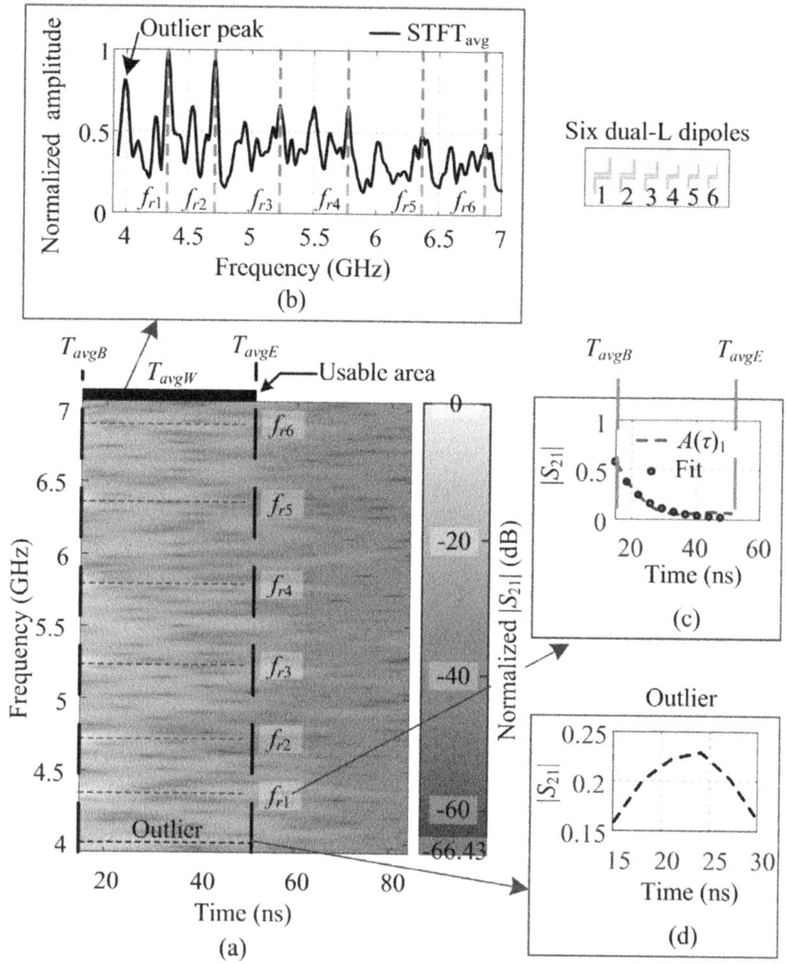

Figure 4.12. *Extraction of CNRs by the spectrogram method for the six dual-L dipoles tag using the uncalibrated raw signal measured in a realistic outdoor environment. (a) STFT calculated using a Hamming window of 30 ns with an overlap of 90%. (b) Extraction of $f_r i$ from the peak apexes of $STFT_{avg}$ signal. (c) Extraction of σ_1 from the damping time signal $A(\tau)_1$ selected at f_{r1} inside T_{avgW}. (d) Detection of the outlier around 4 GHz. For a color version of this figure, see www.iste.co.uk/ali/RFID.zip*

For the extraction of aspect-independent parameters for the multi-scatterer-based tag (see Figure 4.1) using the MPM, the algorithm follows the same parameters as discussed for the results presented in Figure 4.11.

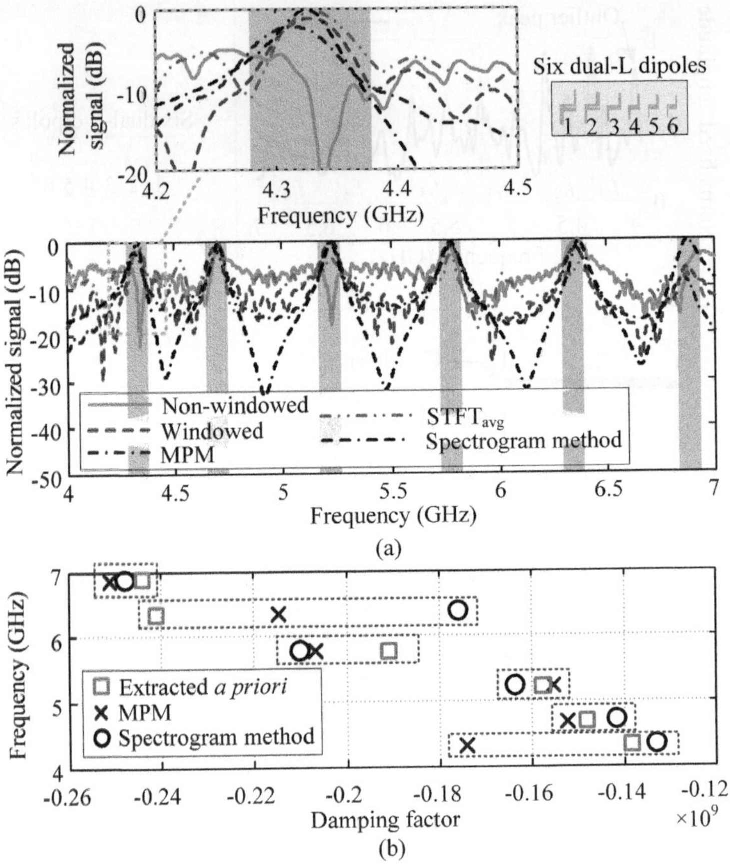

Figure 4.13. *Comparison of the MPM and the spectrogram method. (a) Reconstructed FD responses along with their corresponding supplied raw responses (i.e. raw windowed FD response and raw non-windowed FD response). The raw windowed TD response and raw non-windowed TD response are supplied to the MPM and the spectrogram method, respectively, where they are used to extract the aspect-independent parameters for the six dual-L dipoles tag measured in a realistic outdoor environment. (b) Extracted complex poles in comparison with a priori extracted complex poles. The shaded areas are in the range of* -50 *MHz* $\leq f_{riap} \leq 50$ *MHz. For a color version of this figure, see www.iste.co.uk/ali/RFID.zip*

Figure 4.13(a) shows a comparison of reconstructed FD responses from both the MPM and the spectrogram method along with their corresponding supplied uncalibrated raw responses (i.e. windowed FD response and non-windowed FD response). The uncalibrated raw windowed TD response and uncalibrated raw non-windowed TD response are supplied to the MPM and the spectrogram method, respectively, where they are used to extract the aspect-independent parameters for the six dual-L dipoles tag measured in a realistic outdoor environment. Figure 4.13(b) shows a comparison of extracted complex poles and *a priori* extracted complex pole. The absolute differences of extracted parameters using both the spectrogram method and the MPM from their corresponding *a priori* extracted parameters are outlined in Table 4.3.

Scatterer i	Spectrogram method			Matrix pencil method		
	$\lvert\Delta f_{ri}\rvert$ (MHz)	$\lvert\Delta\sigma_i\rvert \times 10^9$	$\lvert\Delta Q_i\rvert$	$\lvert\Delta f_{ri}\rvert$ (MHz)	$\lvert\Delta\sigma_i\rvert \times 10^9$	$\lvert\Delta Q_i\rvert$
1	≈ 0	0.005	4.01	15.6	0.036	20.54
2	9.77	0.007	4.92	11.75	0.004	2.93
3	9.77	0.006	3.54	8.2	0.002	1.73
4	≈ 0	0.019	8.5	5.34	0.016	7.1
5	29.3	0.065	31.17	11.96	0.026	10.31
6	9.77	0.004	1.39	24.17	0.007	2.79

Table 4.3. *Absolute difference of extracted parameters using the spectrogram method and the MPM from the corresponding* a priori *extracted parameters – realistic outdoor environment*

From Figure 4.13 and Table 4.3, it can be observed that the extraction of the parameters for the fifth scatterer of the six dual-L dipoles tag is worst in the spectrogram method. Otherwise, the spectrogram method provides always very good results. To overcome the detection of the outliers and to improve the accuracy of parameter extraction, we adopted an improvement in the spectrogram approach by using dedicated STFT averaging windows limited in both time and frequency to each scatterer of the employed tag based on *a priori* extracted parameters. Figure 4.14(a) shows the extraction of CNRs by the spectrogram method with multiple dedicated averaging windows for the six dual-L dipoles tag using the uncalibrated raw signal measured in a realistic outdoor environment (i.e. plotted in Figure 4.13, green curve). For all averaging windows, T_{avgB} is fixed at 15 ns and T_{avgE} is calculated using [4.16], with *a priori* information of each scatterer with a threshold amplitude $a_{en} = 0.001$. These dedicated averaging windows are also limited in frequency band $BW_{100\ MHz}$ ranging from – 50 MHz to 50 MHz around f_{riap} of each scatterer. Figure 4.14(b) and (c) shows the extraction of f_{ri} and σ_i using such dedicated averaging windows, for example, for the third scatterer of the six dual-L dipoles tag. The former technique (Figure 4.13) is same as the latter one (Figure 4.14) if it exhibits an averaging window around the maximum of the signal peak apexes limited in the independent time durations for all resonant scatterers.

Figure 4.15 shows the comparison of extracted complex poles for the six dual-L dipoles tag using the spectrogram method with one averaging window in an anechoic environment (presented in Figure 4.11(b)), the spectrogram method with one averaging window in a realistic outdoor environment (presented in Figure 4.13(b)), the spectrogram method with multiple dedicated averaging windows in a realistic outdoor environment (presented in Figure 4.14) and *a priori* extracted complex poles. For the realistic outdoor environment, it is demonstrated that the spectrogram method with multiple dedicated averaging windows has a better agreement with an *a priori* extracted parameter than the spectrogram method with one averaging window.

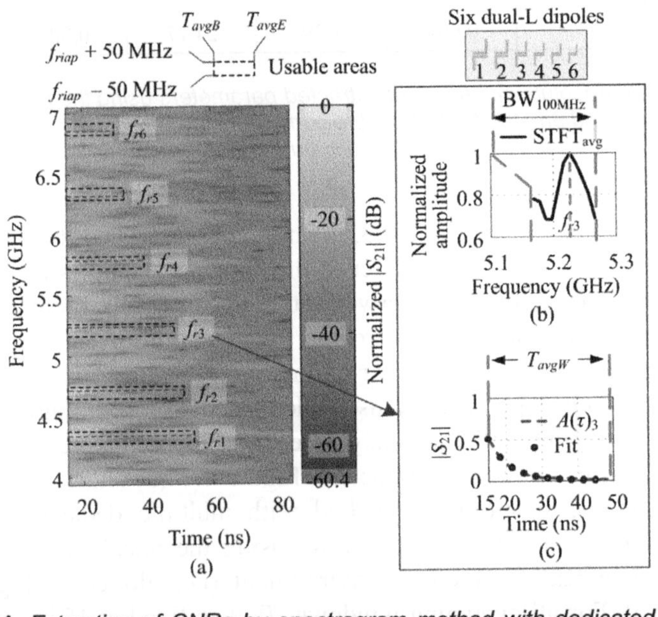

Figure 4.14. *Extraction of CNRs by spectrogram method with dedicated averaging windows for the six dual-L dipoles tag using the uncalibrated raw signal measured in a realistic outdoor environment. (a) STFT calculated using a Hamming window of 30 ns with an overlap of 90%. (b) Extraction of f_{r3} from the peak apexes of $STFT_{avg}$ signal dedicated to the third dual-L dipole. (c) Extraction of σ_3 from the damping time signal $A(\tau)_3$ selected at f_{r3} inside the T_{avgW} dedicated to the third dual-L dipole. For a color version of this figure, see www.iste.co.uk/ali/RFID.zip*

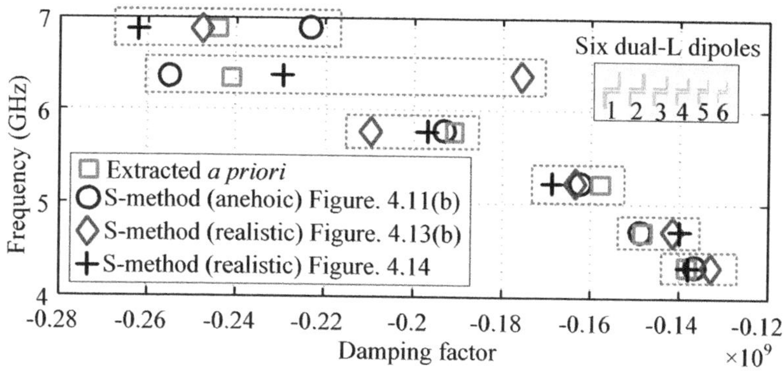

Figure 4.15. *Comparison of extracted complex poles for the six dual-L dipoles tag using the spectrogram method with one averaging window in an anechoic environment (Figure 4.11(b)), the spectrogram method with one averaging window in a realistic outdoor environment (Figure 4.13(b)), the spectrogram method with multiple dedicated averaging windows in a realistic outdoor environment (Figure 4.14) and a priori extracted complex poles. For a color version of this figure, see www.iste.co.uk/ali/RFID.zip*

Scatterer i	**Spectrogram method (one averaging window) (Figures 4.12 and 4.13)**			**Spectrogram method (dedicated averaging windows) (Figure 4.14)**		
	$\|\Delta f_{ri}\|$ **(MHz)**	$\|\Delta\sigma_i\| \times 10^9$	$\|\Delta Q_i\|$	$\|\Delta f_{ri}\|$ **(MHz)**	$\|\Delta\sigma_i\| \times 10^9$	$\|\Delta Q_i\|$
1	≈ 0	0.005	4.01	≈ 0	0	0.19
2	9.77	0.007	4.92	9.77	0.008	6.09
3	9.77	0.006	3.54	9.77	0.011	6.65
4	≈ 0	0.019	8.5	≈ 0	0.006	2.93
5	29.3	0.065	31.17	29.3	0.011	4.53
6	9.77	0.004	1.39	9.77	0.018	6.21

Table 4.4. *Absolute differences of extracted parameters using the spectrogram method with one averaging window and spectrogram method with multiple dedicated averaging windows in realistic outdoor environment from their corresponding* a priori *extracted parameters*

Table 4.4 presents the absolute differences of extracted parameters using the spectrogram method with one averaging window (outlined in Table 4.3) and the spectrogram method with multiple dedicated averaging windows from their corresponding *a priori* extracted parameters. By using the multiple dedicated averaging windows, the extraction accuracy of σ_i is improved, where the value of $|\Delta Q5^{SM}|$ for the fifth scatterer of the six dual-L dipoles tag is improved from 31.17 to 4.53 (see highlights in Table 4.4). The reason behind this improvement is that the multiple dedicated averaging windows provide only the most effective information over time, while in one averaging window, longer time information for the fast damping modes creates ambiguity in the regression process.

4.5. Comparison of computational time durations between the matrix pencil method and the spectrogram method

A comparison of computational time durations has been made between the spectrogram method and the MPM. The single dual-L dipole tag (see inset of Figure 4.1) is used for the experimental results. The measurements are performed in an anechoic environment in a monostatic cross-polarization configuration, as shown in Figure 4.1. Figure 4.16 shows the TD responses for the single dual-L tag calculated from the calibrated FD response measured in an anechoic environment. To calculate the computational time durations, time windowing is applied to measure the signal in various lengths. The beginning time of time window T_{LB} is kept fixed to discard the structural mode, while the ending time of time window T_{LE} is varied to measure different lengths of the signal. The window length from 100 ns to 1000 ns with equal steps of 100 ns has been measured. To calculate the computational time durations, both methods (spectrogram method and MPM) are fed by the same windowed signal (i.e. exhibiting equal length in time duration). The calculated computational time durations for both methods (spectrogram method and MPM) using different windows' lengths are presented in Figure 4.17. This calculation is conducted using MATLAB on an Intel (i7-5600U) processor. The MPM is found to be computationally very expensive compared to the spectrogram method. For a window of 100 ns, the spectrogram method took approximately 4.7 ms, while the MPM took 28.5 ms. For a window of 1000 ns, the spectrogram method took 37.8 ms, while the MPM took 9.15 s.

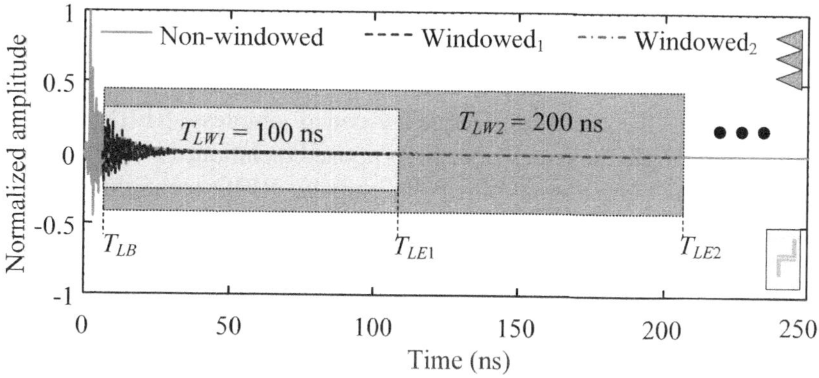

Figure 4.16. *Windowed TD responses with window length T_{LW} ranging from 100 ns to 1,000 ns with a step size of 100 ns for the single dual-L dipole tag calculated from the calibrated FD response measured in an anechoic environment. For a color version of this figure, see www.iste.co.uk/ali/RFID.zip*

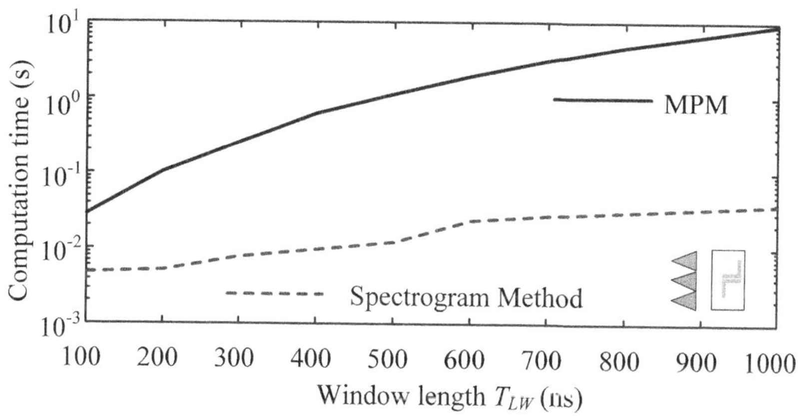

Figure 4.17. *Comparison of computational time durations between the spectrogram method and the MPM. For a color version of this figure, see www.iste.co.uk/ali/RFID.zip*

4.6. Conclusion

Robust detection of the depolarizing REP tags using FFT-based STFT was demonstrated. We have seen that as in the frequency-coded chipless

RFID technology, the resonances of the scatterers are orthogonal to each other, the spectrogram method is an efficient and fast choice. The extraction of complex natural frequency(ies) using the spectrogram has never been done before in the field of frequency-coded chipless RFID. With the operation of a single measurement, the proposed technique is very promising for the practical implementation of the chipless RFID technology, as it is computationally less expensive due to the inherent fast property of FFT. Thus, the proposed technique requires fewer resources and effort. In the context of chipless authentication, aspect-independent parameters for a multi-scatterer tag can accurately be extracted by using the proposed spectrogram approach. These aspect-independent parameters are the main information contributing to the uniqueness of the tags coming from the natural process variations because the parameters are independent of the aspects of the measurement setup.

5

Chipless Authentication Using PCB Tags

5.1. Introduction

In this chapter, a chipless authentication method based on solely naturally occurring randomness (i.e. without any purposely applied variations) in the PCB realization technique is presented. The natural randomness along the geometrical dimensions of a microstrip circuit vitally depends on the technology selected for circuit realization. For chemical etching-based PCB realization technology, non-systematic under- and over-etching are among the fabrication process anomalies (Khandpur 2006, Chap. 9; Pecht and Ganesan 2006) which give rise to the natural process variations up to a slight extent (which is in the order of the metal thickness). For the first time, we proposed that the PCB realizations of simple scatterers can be used as unique footprints. The proposed approach is very low cost and simple because we do not add any particle into the labels as done in (Perret *et al.* 2012). The natural dimensional variations in chipless tags' realizations give rise to the unique variations in the RF field that is compatible with high-level security. We have proved the concept with a sufficiently large population of chipless tags (i.e. 45 tags). For similarity comparison between the PCB realizations, the aspect-dependent part of the measurements has been discarded and merely the signal part containing the aspect-independent parameters (i.e. the frequency of resonance f_r and the quality factor Q) is used. These aspect-independent parameters are the main information possessing the uniqueness of the tags coming from the natural process variations because these parameters are independent of the aspects of the measurement setup (e.g. the position of the tag or the measurement bench in the environment, and the environment itself). Owing to the aspect-independent information, we can simplify the measurement process (e.g. need for fewer resources to store the information in the database) and increase the performance. Next, an in-depth statistical study has been presented to provide an accurate idea of the performance that can be reached. The natural dimensional variations in the realizations of the chipless tags are characterized using microscopic analysis. Then, a relationship between the natural dimensional variations and the similarity variations (that are due to the aspect-independent parameters) is established. Finally, the generalization of the proposed approach is presented. For these reasons, this chapter can be differentiated from the existing work in the literature (Perret *et al.* 2012). This chapter is organized as follows:

– section 5.2 presents the design and the optimization of the chipless RFID tags to be employed for authentication;

– section 5.3 describes the minimum detectable dimensional variations and the tags' discrimination results;

– section 5.4 describes the detection of natural randomness, the authentication results and the generalization of the proposed method;

– finally, section 5.5 draws the conclusions.

As discussed in Chapter 3, the similarity comparisons are performed in the FD using CS [3.1] and in the TD [3.2] using CC_{max} (see Ali *et al.* 2018a, 2019, 2020). From both [3.1] and [3.2] similarity metrics, a result close to a value zero mentions that the two compared signals are not the same and a result close to a value one identifies that the two compared signals are identical.

5.2. Design and the optimization of chipless tags to be employed for authentication

In this section, we have explained that a classical chipless RFID tag is not sensitive enough to be employed for authentication. Therefore, an optimization to enhance the sensitivity of the chipless tag should be done. This fact is explained by simulations and the measurements of the chipless tags exhibiting the purposely applied dimensional variations. Finally, a second bandpass filter model is used to create 2D similarity maps. With the help of the 2D similarity maps, the performance of the classical chipless RFID tags is compared with the performance optimized chipless RFID tags. Here, we have chosen a widespread conventional chemical etching-based

PCB realization technique (which is known and proven) to give an accurate reference of what can be achieved by employing a chipless approach in the authentication. Other realization technologies such as inkjet printing may exhibit higher randomness due to an error in the placement of the ink droplets (Chapter 6). Sensitivity towards naturally occurring randomness during the fabrication process and simplicity in the design of a chipless tag to be employed for authentication application are the key and necessary characteristics. The C-folded dipole is a simple design that can be realized with a lot of different techniques (some of them can be of low cost). This planar scatterer has been extensively used in the field of chipless RFID technology (Perret 2014, Chap. 5). As discussed in Chapter 3, the frequency of resonance f_r of such a resonator can be calculated using [3.5].

5.2.1. *C-folded uni-scatterer tags (classical design)*

Figure 5.1 shows the layout of a simulated chipless tag based on a C-folded uni-scatterer. For simulations, a commercial full-wave simulator (CST Microwave Studio) is used, where the chipless tags are illuminated by a plane wave. The nominal geometrical dimensions of this simulated scatterer are $L = 3.16$ mm, $g = 2.5$ mm and $w = 0.22$ mm, while the overall size of the tag is 8×6 mm^2. Concerning the value of g, it has been discussed in (Vena *et al.* 2011) that the RCS level is increasing almost linearly with g. Therefore, to increase the RCS level, a higher value of $g = 2.5$ mm is taken.

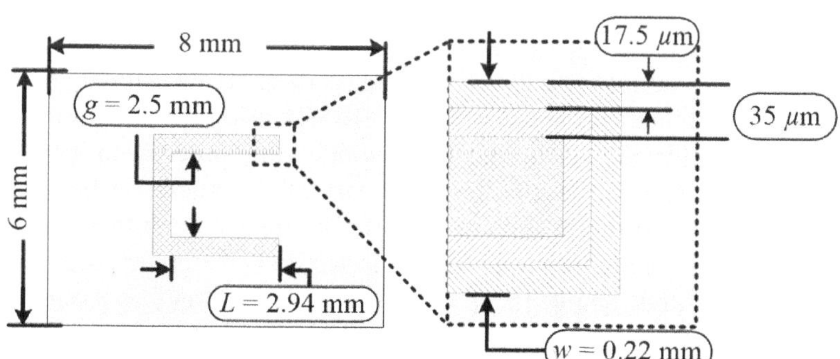

Figure 5.1. *Layout of the simulated C-folded uni-scatterer tag. Inset: the purposely applied variations along the trace are shown in different line patterns. For a color version of this figure, see www.iste.co.uk/ali/RFID.zip*

For nominal dimensional parameters, simulated CST backscattered TD and FD responses are shown in Figure 5.2(a) and (b), respectively. It is known that a backscattered field from a scatterer constitutes two modes: the early part of the signal called the structural mode and the late part of the signal called the antenna mode. Between these two modes, the structural mode is nonessential in our case due to its association with direct optical reflections from the tag and tag's support, while the antenna mode is essential as it is related to reflections due to the resonance of the scatterer. For this reason, time windowing (see section 3.5.1 or section 4.3 for explanation) is applied to extract the middle part of the signal corresponding to the antenna mode (see Figure 5.2(a) and (b)).

The whole backscattered response (comprising both the structural and antenna modes) from a C-folded scatterer can be modeled by a low pass filter analytical model as discussed in (Vena *et al.* 2011). On the other hand, if the backscattered TD response is comprising merely the antenna mode extracted by applying the time windowing to the TD response, a second-order bandpass filter analytical model can also be employed to model such a kind of C-folded scatterer (see Ali *et al.* 2018b):

$$T(\omega) = G\left[\frac{\frac{2mj\omega}{\omega_0}}{1+\frac{2mj\omega}{\omega_0}+\left(\frac{j\omega}{\omega_0}\right)^2}\right], \qquad [5.1]$$

where m and ω_0 correspond to the damping ratio and the angular frequency of resonance, respectively, of a pole of second order. The G is the gain determining the amplitude level of the backscattered signal from the scatterer. To validate this analytical model [5.1], a time window of 3 ns is applied to the TD signal in the time windowing process to extract the antenna mode, where 2 ns of the early part of the signal is neglected to discard the structural mode (Figure 5.2(a)). The resonance frequency ω_0 and the damping ratio m are extracted from the windowed signal. Then, a model signal is generated using [5.1] with extracted m and ω_0 parameters. Subsequently, a fitting comparison among the non-windowed FD signal, the windowed FD signal and the model [5.1] generated signal is shown in Figure 5.1(b). The response of the analytical model [5.1] is in perfect agreement with the simulated windowed FD signal. For the sake of simplicity, the signals are presented in normalized amplitude levels (i.e. $G = 1$). It is important to note that the extracted m and ω_0 parameters are aspect-independent parameters.

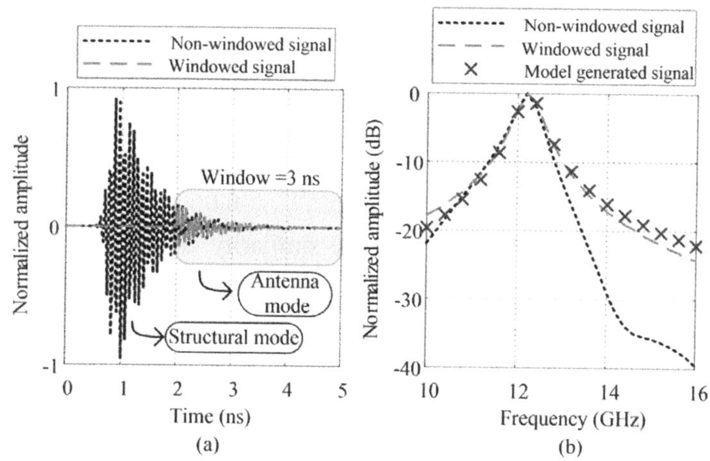

Figure 5.2. *(a) Simulated CST backscattered TD responses. (b) Simulated CST backscattered FD responses. Also, a comparison of the windowed FD signal with a signal generated by the analytical model [5.1]. For a color version of this figure, see www.iste.co.uk/ali/RFID.zip*

In this chapter, a high-performance substrate (Rogers RO4003) is used with $\varepsilon_r = 3.38$ and $h = 0.81$ mm. This substrate is in contrast to the FR4, where ε_r can change from one sample to another. For this reason, the natural changes to be used for authentication are not only linked to the substrate. Initially, to show the sensitivity of a C-folded uni-scatterer chipless tag towards the naturally occurring randomness with the help of simulations, we have purposely applied variations along all sides of the metal trace, as shown in different line patterns (see Figure 5.1). These purposely applied variations are in the order of maximum probable fabrication tolerance, that is, $t = 35$ µm, in the form of a decrease of a step of 17 µm. It is important to note that the application of these purposely applied variations along all sides of the metal trace would affect the w as well as the g *and* L. The sensitivity of this C-folded uni-scatterer tag is presented in Figure 5.3. This sensitivity is calculated by comparing the signals as a result of applied dimensional variations with a reference signal. The reference signal is the one that is obtained at nominal dimensions. Before comparing them, the time windowing is applied to all signals. Next, m and ω_0 parameters are extracted from all simulated windowed FD signals. Then, we generated the model signals by using [5.1] with the extracted m and ω_0 parameters. Finally, the model-generated signals are also compared to calculate the sensitivity of the analytical model [5.1]. The signal similarities (CS and CC_{max}) are calculated by using [3.1] for the FD signals and using [3.2] for the TD signals, as discussed in sections 3.1 and 5.1. As expected, the decrease in similarity occurred because of the applied dimensional variations (see Figure 5.3). At an applied variation of 35 μm, the changes in the similarity metrics [3.1] and [3.2] for both the simulated CST backscattered and the model signals are $\Delta CS_1^{CST} = 0.06$, $\Delta CC_{max1}^{CST} = 0.09$, $\Delta CS_1^{Model} = 0.06$ and $\Delta CS_{max1}^{Model} = 0.06$, respectively. It can be observed that the similarity change calculated using all data points of the simulated CST backscattered signals (ΔCS_1^{CST} and ΔCC_{max1}^{CST}) and the similarity change calculated using the model [5.1] signals (ΔCS_1^{Model} and ΔCS_{max1}^{Model} that are generated by using merely two aspect-independent m and ω_0 parameters) are almost equal. It is clear that the significant information in the backscattered signals is related to m and ω_0 parameters due to their aspect-independent nature.

The quality factor Q is related to the damping ratio m: $Q = 1/(2m)$. The quality factor Q of such a C-folded theoretical scatterer, with a very small value of g, can also be calculated using the following expression (Gov and Shtrikman 1995):

$$\frac{1}{Q} \simeq \left(\frac{g}{\lambda}\right)^2. \qquad [5.2]$$

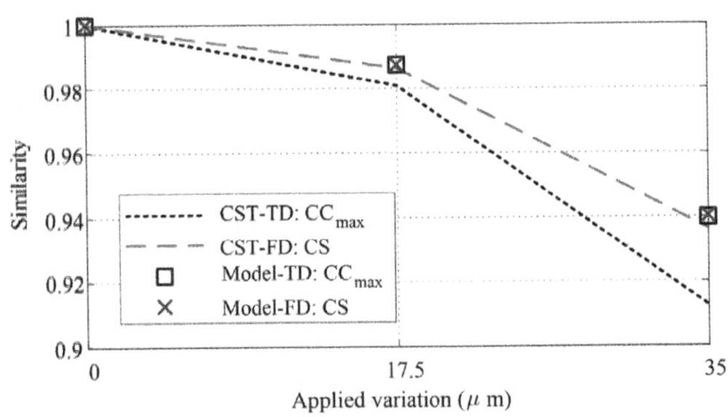

Figure 5.3. *Similarity level calculated in both FD [3.1] and TD [3.2] using the simulated CST backscattered responses as well as the model generated signals. At an applied variation of 35 μm, the changes in the similarity metrics [3.1] and [3.2] for both the simulated CST backscattered and model signals are* $\Delta CS_1^{CST} = 0.06$, $\Delta CC_{max1}^{CST} = 0.09$, $\Delta CS_1^{Model} = 0.06$ *and* $\Delta CC_{max1}^{Model} = 0.06$, *respectively. For a color version of this figure, see www.iste.co.uk/ali/RFID.zip*

To practically validate the sensitivity of such C-folded uni-scatterer tags by experimental results, we have realized two groups of prototypes, where each group is constituted of five tags. The group N exhibits nominal dimensional parameters exactly the same as the nominal dimensional parameters of the simulated tag, as depicted in Figure 5.1. To apply a variation on the subsequent group, we shortened the L by 100 μm for group R100. The purpose behind this large applied variation (as compared to 35 μm) is to clearly discriminate the chipless tags without any ambiguity (i.e. observational noise or the systematic noise). A photograph of the first tag from each group along with their dimensions is shown in Figure 5.4. The overall size of these fabricated tags is also the same as the simulated ones: 8×6 mm^2.

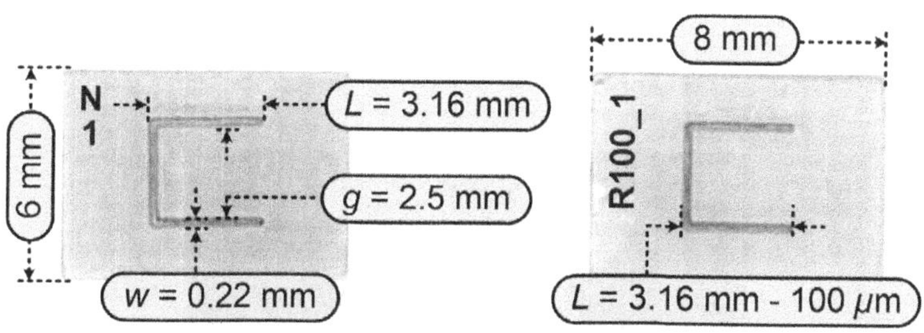

Figure 5.4. *Top view of the fabricated C-folded uni-scatterer tags along with their dimensions. For a color version of this figure, see www.iste.co.uk/ali/RFID.zip*

The measurement setup for the C-folded uni-scatterer tags is shown in Figure 5.5, where the measurements are done in a bistatic co-polarization configuration in an anechoic environment. The experimental results are measured using a VNA (Agilent Technologies PNA 5222A) with a source power of –5 dBm. The SATIMO QH2000 (2–32 GHz) horn antennas are used. The distance between the antennas and the tag is $r = 16.5$ cm, while the distance between both antennas is $e = 2.7$ cm. The measured quantity is the transmission coefficient S_{21}, as shown in Figure 5.6. For each tag, five repetitive measurements are taken. For each measurement trial, the tag is removed and repositioned at its position to ensure the validity of the repeatability in the presence of an observational error. Furthermore, background normalization is applied (i.e. the subtraction of an instantaneous measurement taken in the absence of the tag from the measurement taken in the presence of the tag). It can be observed that the measured non-windowed signal (see Figure 5.6(b)) does not exhibit a structural mode as compared to the simulated CST backscattered signal (see Figure 5.2(a)). The reason behind this is that measurements are made using bistatic co-polarization configuration, while in simulations an E-field probe is used (which is similar to monostatic co-polarization configuration). In such a case (bistatic measurement configuration), a significant part of the structural mode is already removed due to the nature of the configuration. So, the time windowing procedure (see section 3.5.1 or section 4.3 for explanation) is not needed to discard the structural mode. To explain this fact, a time window of 10 ns is employed to extract the antenna mode, where 1.6 ns of the early part of the signal is discarded to neglect the minor structural mode (Figure 5.6(b)). In Figure 5.6(a), it can be observed that both FD non-windowed and FD windowed signals do not differ a lot from each other, in contrary to the simulated CST backscattered FD signals in Figure 5.2(b).

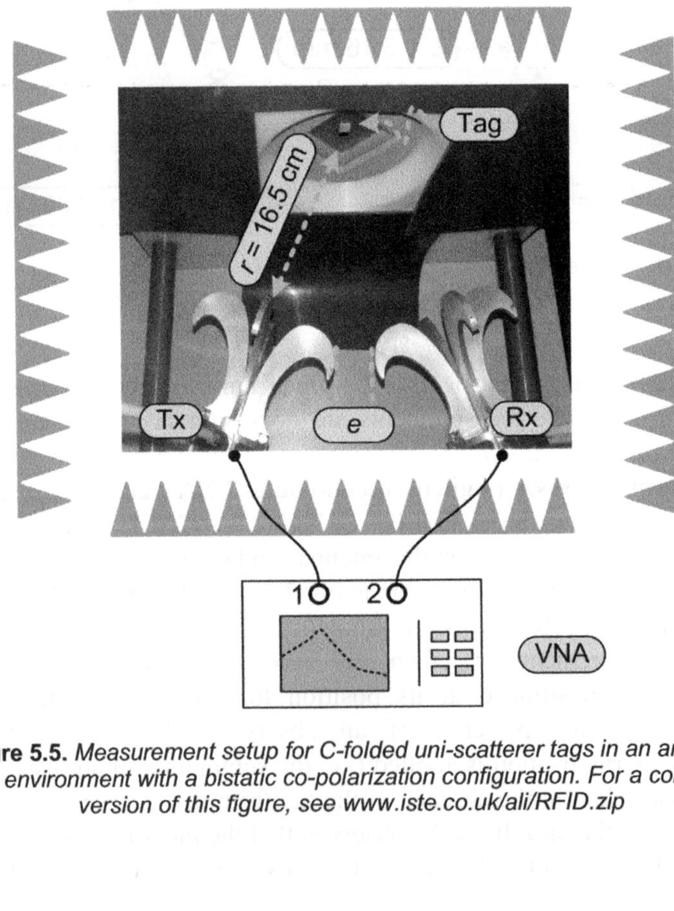

Figure 5.5. *Measurement setup for C-folded uni-scatterer tags in an anechoic environment with a bistatic co-polarization configuration. For a color version of this figure, see www.iste.co.uk/ali/RFID.zip*

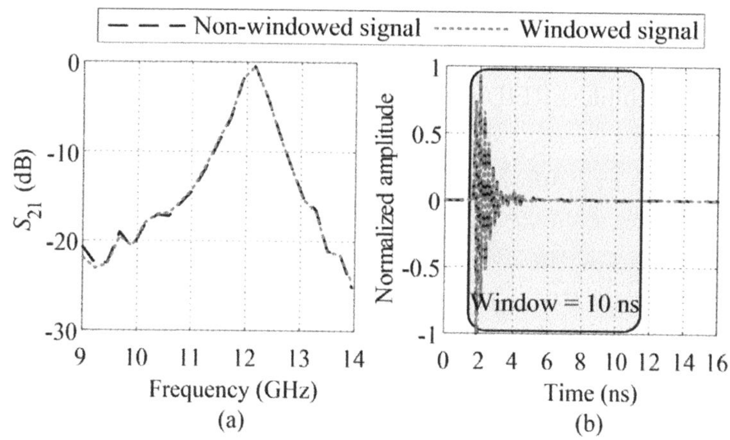

Figure 5.6. *Time windowing of the transmission coefficient S_{21} of the C-folded uni-scatterer tag N1. (a) Frequency domain responses. (b) Time domain responses. For a color version of this figure, see www.iste.co.uk/ali/RFID.zip*

A CS map for uni-scatterer tags is presented in Figure 5.7. To construct this map, we extracted the f_r and Q parameters of all 25 measurements related to each group. The mean values of the extracted f_r and Q parameters of group N, $f_r^m = 12.05$ GHz and $Q^m = 27$ are taken as a reference. First, a reference signal x^r is generated by using the analytical model [5.1] with the reference f_r^m and Q^m parameters. Then, a varying signal x^v is generated using the analytical model [5.1] by varying the f_r^m and Q^m parameters such that $f_r^v = f_r^m - 0.2$ GHz to $f_r^v = f_r^m + 0.6$ GHz and $Q^v = Q^m - 10$ to $Q^v = Q^m + 50$. Subsequently, the reference signal x^r and the varying signal x^v are compared in the form of complex FD signals using CS [3.1]. Finally, the CS score is plotted versus Δf_r and ΔQ such that $\Delta f_r = f_r^v - f_r^m$ and $\Delta Q = Q^v - Q^m$. Furthermore, the extracted f_r and Q parameters related to all 25 measurements of each group are superimposed on the similarity map for an intuitive comparison. The centroid to centroid difference of CS between the groups N and R100 is $\Delta CS_1 = 0.13$. This $\Delta CS_1 = 0.13$ is large as compared to the simulated $\Delta CS_1^{CST} = 0.06$ (see Figure 5.3). The reason for the difference between ΔCS_1 and ΔCS_1^{CST} is that in simulations, a uniform dimensional variation of 35 μm is applied along all sides of the metal trace, while in measurements, a dimensional variation of 100 μm is applied merely along L. It is important to mention that the natural dimensional variations happened during the realization of uni-scatterer tags are considered as a part of applied dimensional variations. Figure 5.7 shows a useful representation that can provide direct information about the potential performance (e.g. the slope of the CS) of the chipless RFID technology for authentication.

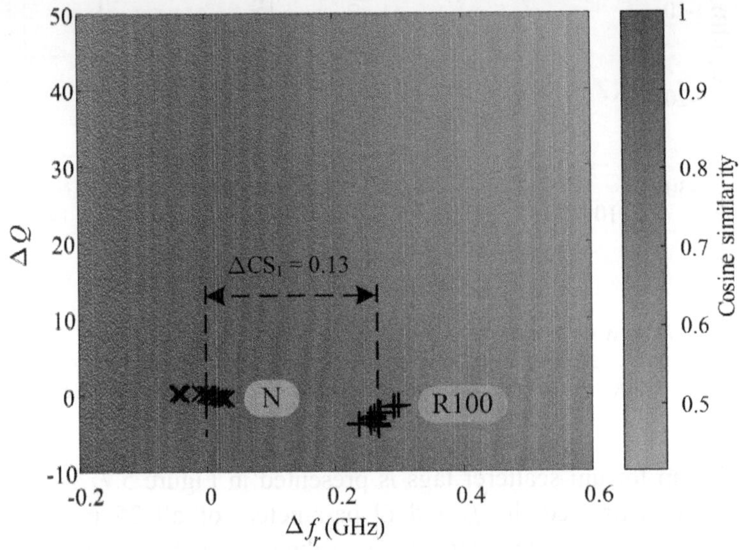

Figure 5.7. *Cosine similarity map obtained by comparing a reference complex FD signal x^r with a varying complex FD signal x^v using CS [3.1] for uni-scatterer tags. The superimposed markers × and + correspond to the extracted parameters of 25 measurements of five tags of N and R100 groups, respectively. For a color version of this figure, see www.iste.co.uk/ali/RFID.zip*

5.2.2. C-folded quad-scatterer tags (optimized design)

From a practical viewpoint, the C-folded uni-scatterer with a low Q factor and a low RCS level of the metallic resonator requires a bistatic measurement apparatus with two antennas. Therefore, we have coupled four scatterers to realize one tag with a higher Q to make it appropriate in practical application. The influence of a high Q chipless tag in authentication is discussed next. Figure 5.8 shows the layout of a simulated chipless tag based on C-folded quad-scatterers. In this quad-scatterer design, the nominal geometrical dimensions of each C-folded scatterer are exactly the same as the geometrical dimensions of the C-folded scatterer of the uni-scatterer tag (see Figure 5.4), while the spacing among the scatterers is s = 0.15 mm and the overall size of the tag is 8 × 15 mm^2.

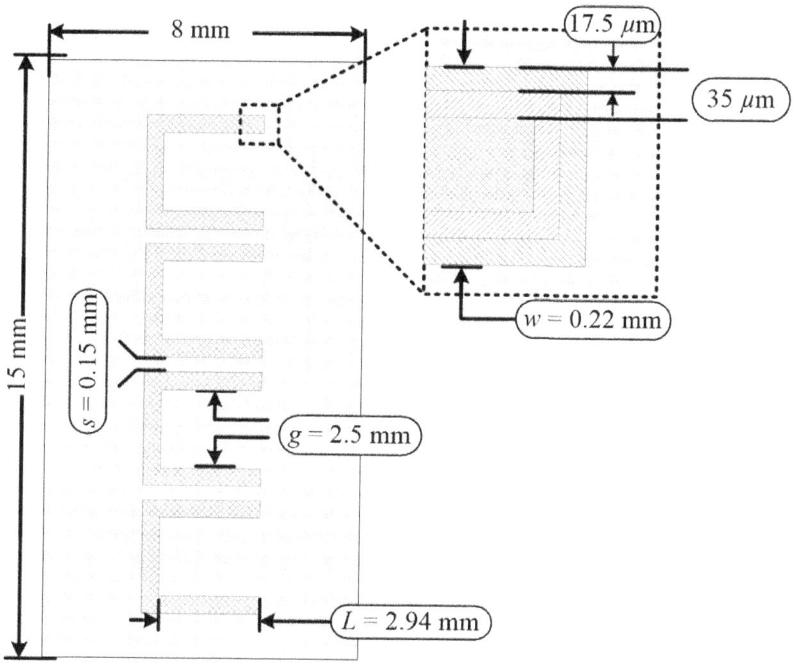

Figure 5.8. *Layout of the simulated C-folded quad-scatterer tags. Inset: the purposely applied variations on the trace are shown in different line patterns. For a color version of this figure, see www.iste.co.uk/ali/RFID.zip*

For the nominal dimensional parameters, the simulated CST backscattered TD and FD responses are shown in Figure 5.9(a) and (b), respectively. The time windowing is achieved by using a time window of 10 ns to extract the antenna mode, while 2 ns of the early part of the signal is neglected to discard the structural mode. From the windowed signals (Figure 5.9(a) and (b)), the ω_0 and m are extracted to generate the model signal from the analytical model [5.1]. Subsequently, a fitting comparison between the windowed FD signal and a model [5.1] generated signal is shown in Figure 5.9(b). It can be observed that the response of the analytical model is in perfect agreement with the simulated windowed FD signal.

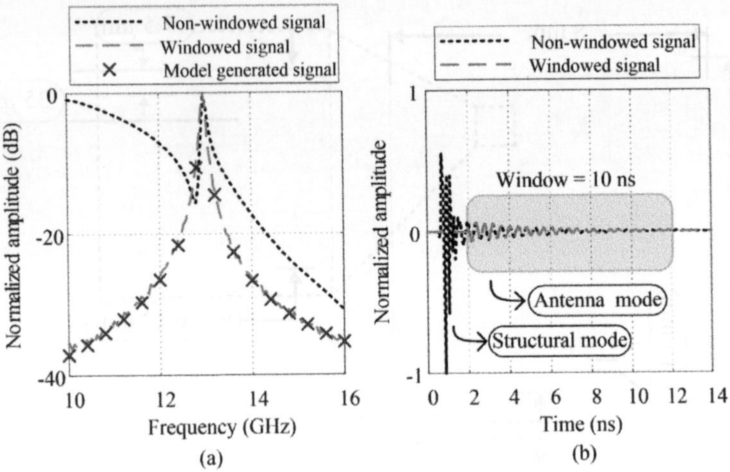

Figure 5.9. *(a) The simulated CST backscattered TD responses. (b) The simulated CST backscattered FD responses. Also, a comparison of the windowed FD signal with a signal generated by the analytical model [5.1]. For a color version of this figure, see www.iste.co.uk/ali/RFID.zip*

Like the uni-scatterer tag as discussed above (see Figure 5.1), we have also applied variations purposely along all sides of the metal trace shown in different line patterns (see Figure 5.8) to check the sensitivity of such C-folded quad-scatterer based tag towards the applied geometrical variations (i.e. in the order of fabrication tolerance) in simulations. The simulated CST backscattered signals and the model-generated signals are compared (as discussed above for Figure 5.3) in FD [3.1] and TD [3.2] to calculate the sensitivity of quad-scatterer tags, as shown in Figure 5.10. At an applied variation of 35 μm, the changes in the similarity metrics [3.1] and [3.2] for both the simulated CST backscattered and the model signals are $\Delta CS_4^{CST} = 0.22$, $\Delta CC_{max4}^{CST} = 0.23$, $\Delta CS_4^{Model} = 0.23$ and $\Delta CC_{max4}^{Model} = 0.23$, respectively. It can be observed that the similarity change calculated using all data points of the CST backscattered signals (ΔCS_4^{CST} and ΔCC_{max4}^{CST}) and the similarity change calculated using the model [5.1] signals (ΔCS_4^{Model} and ΔCC_{max4}^{Model} that are generated by using merely two aspect-independent m and ω_0 parameters) are almost equal. Again, this behavior highlights the significance of the aspect-independent m and ω_0 parameters.

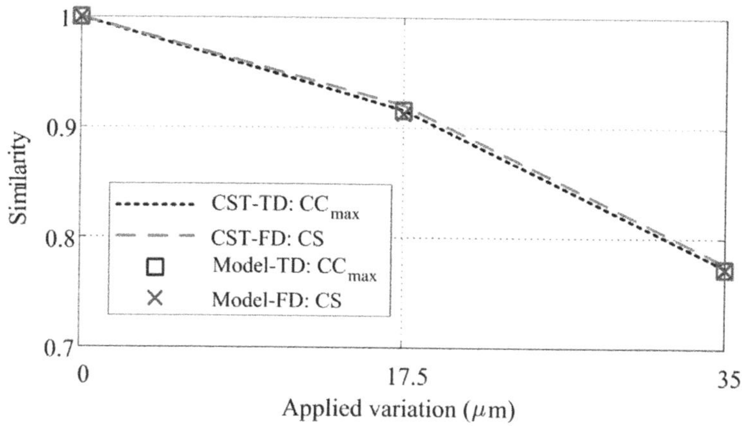

Figure 5.10. *Similarity levels calculated in both FD [3.1] and TD [3.2] using simulated CST backscattered responses and also the model generated signals. At an applied variation of 35 μm, the changes in the similarity metrics [3.1] and [3.2] for both CST backscattered and model signals are* ΔCS_4^{CST} *= 0.22,* $\Delta CC_{\max 4}^{CST}$ *= 0.23,* ΔCS_4^{Model} *= 0.23 and* $\Delta CC_{\max 4}^{Model}$ *= 0.23, respectively. For a color version of this figure, see www.iste.co.uk/ali/RFID.zip*

Similar to the C-folded uni-scatterer tags, the sensitivity of such C-folded quad-scatterer tags is demonstrated by experimental results. For this purpose, we fabricated two groups of prototypes for quad-scatterer tags, where each group is constituted of five tags. The group N exhibits nominal dimensional parameters exactly the same as the simulated tag, as depicted in Figure 5.8. Like the case of the uni-scatterer tag (Figure 5.4), we reduced the L by 100 μm for group R_100. A photograph of all the tags of both groups along with their dimensions is shown in Figure 5.11. The overall size of these fabricated tags is also the same as of the simulated ones: 8 × 15 mm^2.

The measurement setup for the C-folded quad-scatterer tags is shown in Figure 5.12, where the measurements are done in a mono cross-polarization configuration in an anechoic environment. In this configuration, the test tag is placed at an inclination of 45°. The experimental results are measured using a VNA (Agilent Technologies PNA 5222A) with a source power of −5 dBm. The SATIMO QH2000 (2–32 GHz) horn antenna is used. The distance between the antenna and the tag is $r = 13.5$ cm.

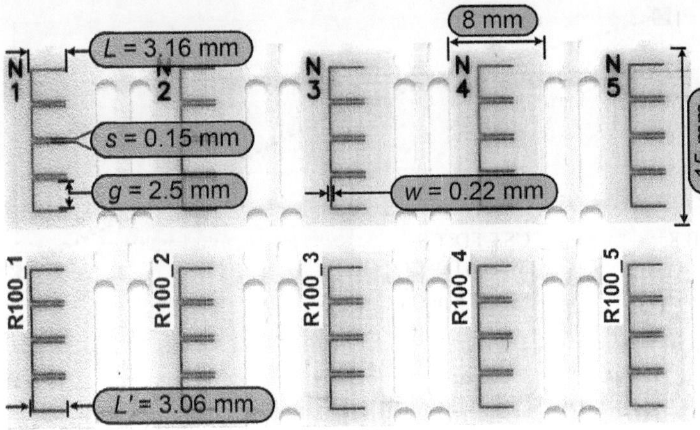

Figure 5.11. *Top view of the fabricated C-folded quad-scatterer tags along with their dimensions. For a color version of this figure, see www.iste.co.uk/ali/RFID.zip*

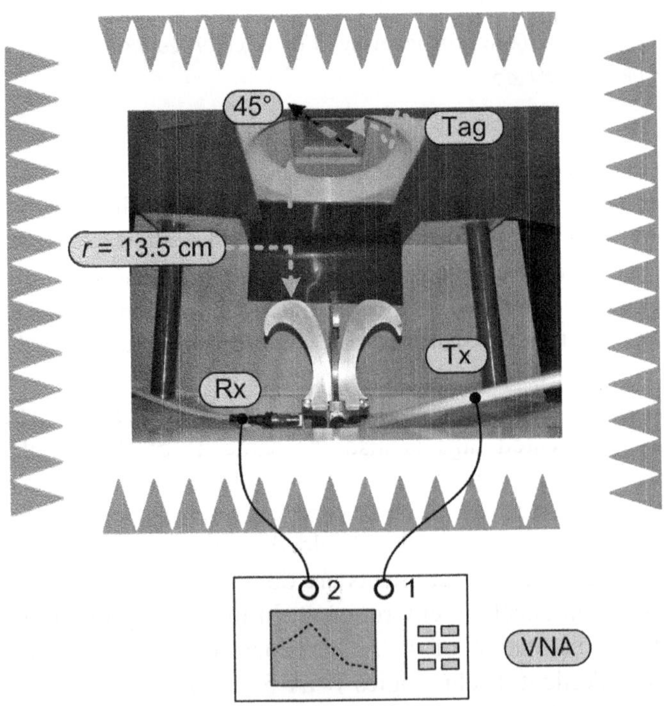

Figure 5.12. *Measurement setup for the C-folded quad-scatterer tags in a cross-polarization configuration in the anechoic environment and in a realistic outdoor environment. For a color version of this figure, see www.iste.co.uk/ali/RFID.zip*

The measured quantity is the transmission coefficient S_{21}, as shown in Figure 5.13. Similar to the C-folded uni-scatterer tags, five repetitive measurements are taken for each C-folded quad-scatterer tag. For each measurement trial, the tag is removed and repositioned at its position to ensure the validity of the repeatability in the presence of an observational error. Furthermore, background normalization is applied (i.e. the subtraction of an instantaneous measurement taken in the absence of the tag from the measurement taken in the presence of the tag). Then, time windowing (see section 3.5.1 or section 4.3 for explanation) with a time window of 10 ns is employed to extract the antenna mode, where 2.5 ns of the early part of the signal is discarded to neglect the structural mode (Figure 5.13(b)).

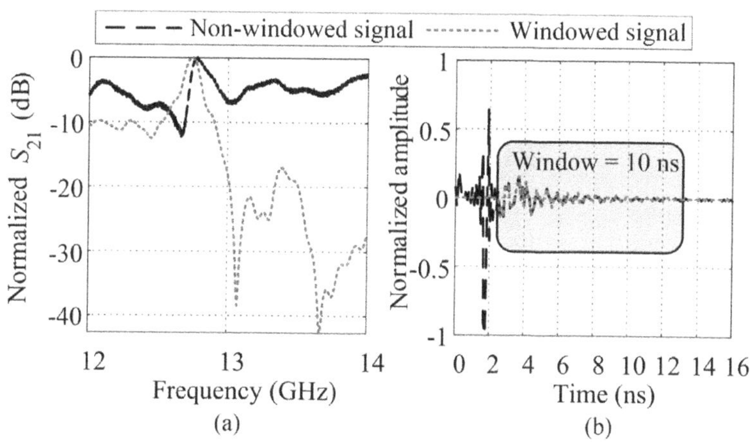

Figure 5.13. *Time windowing of the transmission coefficient S_{21} of the C-folded quad-scatterer tag N1 measured in an anechoic environment. (a) Frequency domain responses. (b) Time domain responses. For a color version of this figure, see www.iste.co.uk/ali/RFID.zip*

To highlight the potential of the C-folded quad-scatterer tags for authentication, a CS map is presented in Figure 5.14. The procedure to calculate the CS map for quad-scatterer tags is the same as discussed above for the CS map for uni-scatterer tags in Figure 5.7. In this case, the calculated reference parameters are $f_r^m = 12.68$ GHz and $Q^m = 98.8$ (i.e. the mean values of the extracted f_r and Q parameters of group N). The centroid to centroid difference of CS between the groups N and R100 is $\Delta CS_4 = 0.6$. This $\Delta CS_4 = 0.6$ is very large than the $\Delta CS_1 = 0.13$ (see Figure 5.7). Also, for this quad-scatterer case (Figure 5.14), the slope of CS is quite sharp along the Δf_r as compared to the uni-scatterer case (Figure 5.7). This sharpness is due to the increase in the Q of the tags from the uni-scatterer tag ($Q^m = 27$) to quad-scatterer tag ($Q^m = 98.8$). Such a large similarity change is the reason for the choice of these quad-scatterer tags to be employed for authentication.

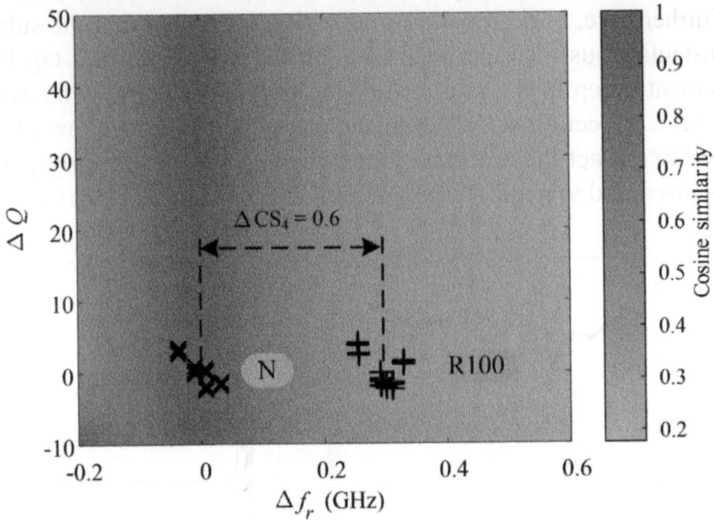

Figure 5.14. *Cosine similarity map obtained by comparing a reference complex FD signal x^r with a varying complex FD signal x^v using CS [3.1] for quad-scatterer tags. The superimposed markers × and + correspond to the extracted parameters of 25 measurements of five tags of N and R100 groups, respectively. For a color version of this figure, see www.iste.co.uk/ali/RFID.zip*

5.3. Detection of minimum dimensional variation in outdoor realistic environment and authentication results

To test the ability of the system to detect these random variations induced by fabrication inaccuracy, an initial work consists of voluntarily modifying the dimensions of the C-folded scatterer to quantify the minimum variations observable. For this purpose, three groups of five identical chipless quad C-folded tags referenced as N, R20 and R50 are realized. The group N exhibits the nominal geometrical dimensions. We purposely shortened the length of the arms L' by 20 µm for R20 group and by 50 µm for R50 group. The reason behind such purposely applied variations is to ensure dimensional changes in fabricated prototypes, where the actual randomness happening in the fabrication process is considered as a part of the purposely applied variations. A photograph of all the tags of three groups and their dimensions are shown in Figure 5.15.

The measurements are done in a monostatic cross-polarization configuration in a realistic outdoor environment by placing the tag at an angle of 45° to generate a cross response. The rest of the measurement parameters such as the antenna, the VNA, the output power and the distance from the antenna to tag are exactly the same as discussed in the measurement setup shown in Figure 5.12. All five tags from each group are successively measured five times, where each measurement is done by removing and repositioning the tag at its position. Furthermore, background normalization is carried out by subtracting an instantaneous measurement taken in the absence of the tag from the measurement taken in the presence of the tag.

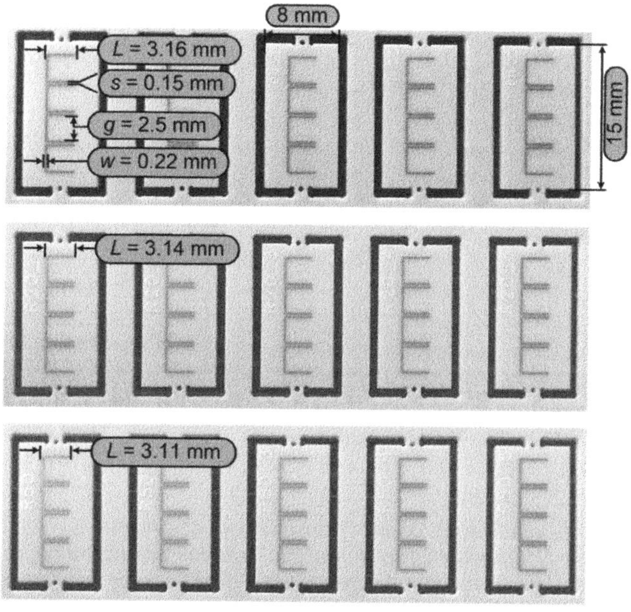

Figure 5.15. *Top view of the fabricated C-folded quad-scatterer tags along with their dimensions for testing the minimum detectable dimensional variation in a realistic outdoor environment. For a color version of this figure, see www.iste.co.uk/ali/RFID.zip*

Figure 5.16 shows the measured S_{21} of the first tag of group N measured in an outdoor realistic environment. The time windowing (see section 3.5.1 or section 4.3 for explanation) is employed to extract the antenna mode with a time window of 10 ns, where 2.2 ns of the early part of the signal is discarded to neglect the structural mode. Figure 5.16 exhibits a good agreement between the windowed FD response (i.e. the antenna mode of the C-folded scatterer) and the FD response generated by the analytical model [5.1]. For the sake of simplicity, the responses are normalized ($G = 1$). This result validates the proposed model, the experimental acquisition, and the time windowing processes, which are systematically performed for the calculation of the Q factor and the resonance frequency f_r of the resonant tag. For example, the extracted values for the measured response shown in Figure 5.16 are Q_m=92.72 and f_{rm}=12.69 GHz. The damping ratio $m = 1/(2Q_m) \approx 5.39\times10^{-3}$.

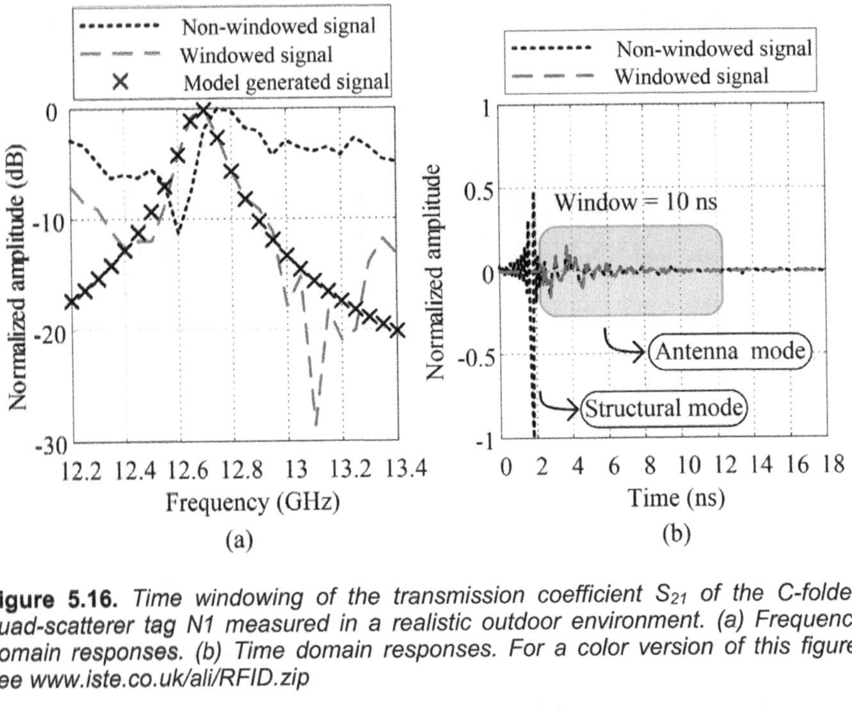

Figure 5.16. *Time windowing of the transmission coefficient S_{21} of the C-folded quad-scatterer tag N1 measured in a realistic outdoor environment. (a) Frequency domain responses. (b) Time domain responses. For a color version of this figure, see www.iste.co.uk/ali/RFID.zip*

A CS map is presented in Figure 5.17. The procedure to calculate the CS map for quad-scatterer tags is the same as discussed above for the CS map for uni-scatterer tags in Figure 5.7. The difference between the CS map presented in Figure 5.14 and the CS map presented in Figure 5.17 is that the former is for the quad-scatterer tags measured in the anechoic environment and the latter is for the quad-scatterer tags measured in a realistic outdoor environment. For the current case (Figure 5.17), the calculated reference parameters are $f_r^m = 12.69$ GHz and $Q^m = 92.72$ (i.e. mean values of the extracted f_r and Q parameters of group N). Furthermore, to demonstrate that the purposely applied variations among all groups N, R20 and R50 would be notable to discriminate them from one to another, the extracted f_{rm} and Q_m of all 25 measurements of each group are superimposed for comparison (Figure 5.17). The mean-to-mean variations of the extracted f_{rm} and Q_m between the groups N and R20 are $\Delta f_{rm(NR20)} = 85.3$ MHz and $\Delta Q_{m(NR20)} = 9.4$, respectively. These variations produce a change in cosine similarity $\Delta CS_{NR20} = 0.17$. In the same manner, the mean-to-mean variations of the extracted f_{rm} and Q_m between the groups R20 and R50 are $\Delta f_{rm(R20R50)} = 86.6$ MHz and $\Delta Q_{m(R20R50)} = 0.6$, which involve a change in cosine similarity $\Delta CS_{R20R50} = 0.18$.

To confirm the existence of the dimensional variations along the L' in the fabricated prototypes, we carefully measured the L' of each arm of each tag by digital microscope images. The measurement setup of a digital microscope is shown in Figure 5.18. A Mighty Scope 5M digital microscope by Aven tools is used, and the calibration of the system is done by the USAF 1951 resolution test chart.

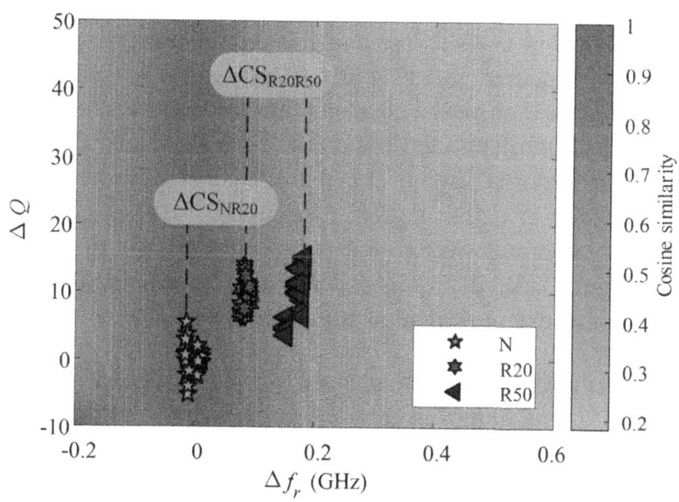

Figure 5.17. *Cosine similarity map obtained by comparing a reference complex FD signal x^r with a varying complex FD signal x^v using CS [3.1] for quad-scatterer tags. The superimposed markers ✫, ✡ and ◁ correspond to the extracted parameters of 25 measurements of five tags of N, R20 and R50 groups, respectively. For a color version of this figure, see www.iste.co.uk/ali/RFID.zip*

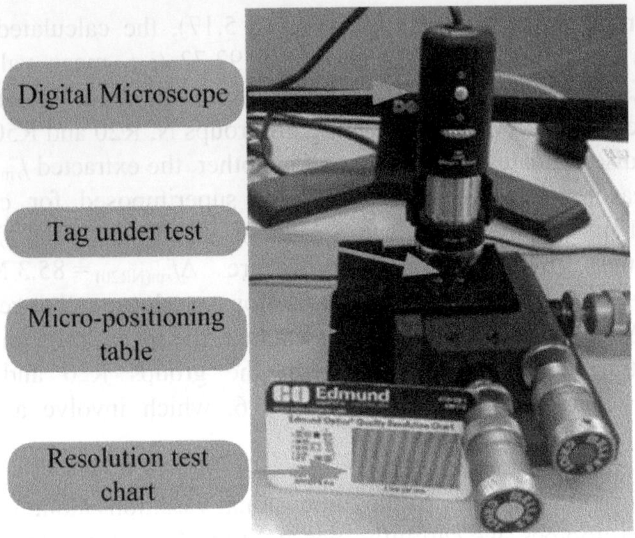

Figure 5.18. *Measurement setup of a digital microscope. Also, the USAF 1951 resolution test chart is shown. For a color version of this figure, see www.iste.co.uk/ali/RFID.zip*

The resolution of the system is approximately 3 µm per pixel. The edge detection technique is used to process the digital images, and an edge to edge distance is measured for each arm of all tags. The coordinates of the measurement line are visually chosen with the mouse cursor. A sample image of the measurements and a sample calibration image are shown in Figure 5.19. In the zoom image of the beginning part of the upper arm of the first C-folded scatterer of the tag N1 (see Figure 5.19), it can be observed that there is an emergence of two edges that may be due to the shadows in the digital image. To keep the consistency, we have selected the midpoint between these two edges to calculate all the distances.

Figure 5.20 shows the measured arms' length L'_m in comparison to the actual value L' for all three groups, where the contribution of the natural randomness can be observed. Since all tags were fabricated on one piece of the substrate, the effect of randomness is correlated among the tags. For all three groups (N, R20 and R50), the mean values of the L'_m distributions are 3.156 mm, 3.134 mm and 3.106 mm, respectively, while the actual values L' for all three groups (N, R20 and R50) are 3.16 mm, 3.14 mm and 3.11 mm, respectively. It is clear that for all three groups the mean values of the L'_m

distributions are less than their corresponding expected values L'. According to Figure 5.20, the margin between the L'_{m} distributions of N and R20 is 3.2 μm, while the margin between the L'_{m} distributions of R20 and R50 is 8.3 μm.

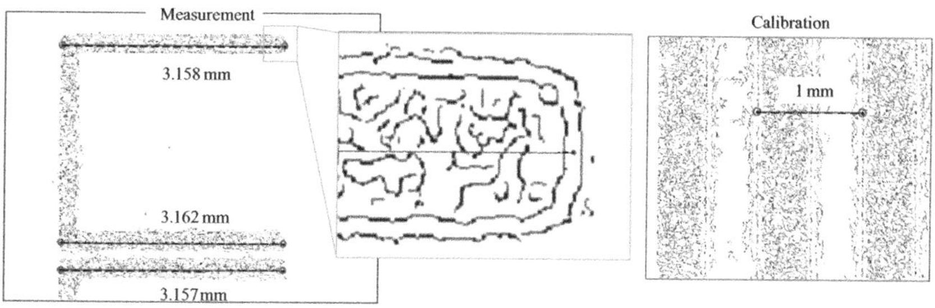

Figure 5.19. *A sample image of the measurement of dimensions. Also, an edge image of the beginning part of the upper arm of the first C-folded scatterer of the first tag of group N. At right, a sample calibration image. For a color version of this figure, see www.iste.co.uk/ali/RFID.zip*

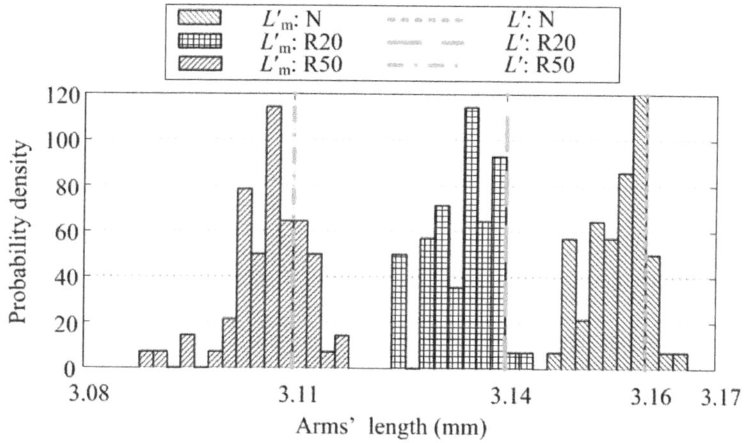

Figure 5.20. *Measured arms' length L'_m for all groups in comparison to the theoretical arms' length L'. For a color version of this figure, see www.iste.co.uk/ali/RFID.zip*

The signals from the two groups are compared among them. The signals' comparison conventions for two distinct groups are outlined in Figure 3.11.

Considering one group, for instance, the group N, the total number of measurements is 25 (each group consists of five tags measured five times each). Therefore, we deduced the number of inter-group combinations among two different groups: $C_2^{50} - 2(C_2^{25}) = 625$. For intra-group, it is found that the total number of combinations to compare all measurements within a specific group is $C_2^{25} = 300$. For all possible combinations, the measurements are compared in the form of complex windowed FD signals using CS [3.1]. Figure 5.21 illustrates the intra-group and inter-group CS distributions. The margin between the maximum inter-group CS and the minimum intra-group CS is about 0.065. The fitted Gaussian probability density functions are used to calculate the PFP and the PFN. The procedure for the calculation of the PFP and PFN is similar to the procedure explained in section 3.4. In this regard, the reader can imagine intra- and inter-group distributions in place of intra- and inter-tag distributions in Figure 3.3(b), respectively. We found a PE lower than 10^{-3}, when the closest distributions are chosen (i.e. worst case): inter-group R20 versus R50 distribution and intra-group R50 distribution (see the inset of Figure 5.21). A comparison between the measurements and the model [5.1] is also presented in Figure 5.21. To calculate the model results, the average values of extracted f_{rm} and Q_m of all 25 measurements of each group are used to generate a complex FD signal, and subsequent signals are compared among the groups. The agreement of the theoretical results with the experimental ones is good. It shows that it is possible to discriminate the tags in a realistic environment merely by focusing on independent quantities f_{rm} and Q_m.

5.4. Detection of natural randomness and authentication results

To demonstrate the authentication based solely on naturally occurring randomness in the fabrication process, we have realized chipless tags based on C-folded quad-scatterers exactly exhibiting the same geometrical dimensions (i.e. without any purposely applied variations) as the geometrical dimensions of group N depicted in Figure 5.11. These tags are realized two times intermittently, where each realization consists of 45 tags. The two different realizations share the same company, same technology, but a different film mask to ensure the natural dimensional randomness.

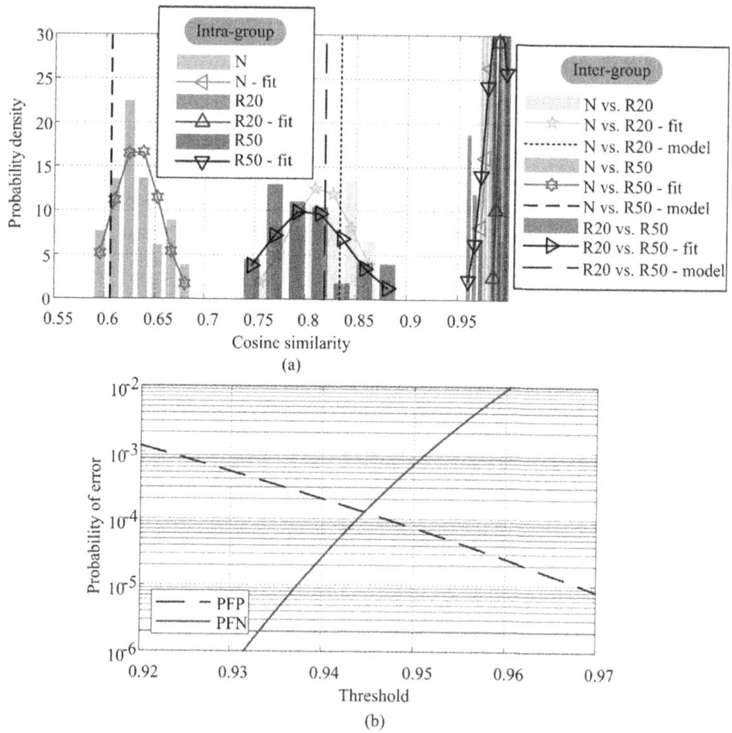

Figure 5.21. *(a) Intra-group and inter-group cosine similarity distributions. Also, a comparison between the theoretical model and the experimental results. (b) Error rates between inter-group R20 versus R50 distribution and intra-group R50 distribution. For a color version of this figure, see www.iste.co.uk/ali/RFID.zip*

Figure 5.22 shows the top view of 45 tags of the first realization exhibiting natural randomness. The f_r of all these tags is an average value of approximately 12.68 GHz, as shown in Figure 5.13.

For the rest of this chapter, the measurement setup and the post-processing (i.e. background normalization and time windowing) of the measured signals (i.e. S_{21}) are also exactly the same as shown in Figures 5.8 and 5.9. In addition, the time windowing process is one step further improved by applying the frequency windowing on the FD windowed signal. This frequency windowing is done by keeping the 200 points around the peak apex while discarding all other points. This procedure is adopted to consider only the aspect-independent parameters (f_r and Q) that principally reside in the top part of skirts of the peak apex. Hence, the FD windowed signal (i.e. also subjected to the frequency windowing) consists of 200 points. For the TD windowed signal, if we ignore the zero padded points, the signal consists of 101 points. The zero padded points can be ignored because they do not contribute to the similarity change. Instead, the zero padding in the TD signal was done only to improve the resolution of the FD windowed signal while computing the inverse Fourier transform in the time windowing process. For the rest of this chapter, we take into account the windowed signals (i.e. also subjected to the frequency windowing) for the similarity analysis.

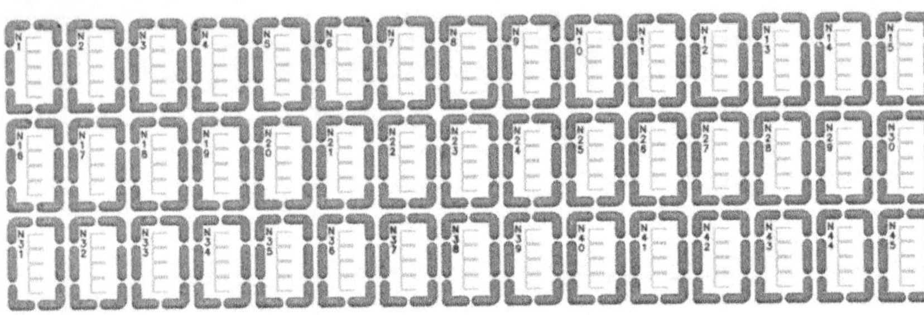

Figure 5.22. *Top view of the first realization of C-folded quad-scatterer tags exhibiting natural randomness. For a color version of this figure, see www.iste.co.uk/ali/RFID.zip*

The two realizations are compared within each realization and also between them (see signals' comparison conventions in Figure 3.3).

5.4.1. *Authentication within each realization*

First, authentication outcomes based on solely naturally occurring randomness within each realization for both intermittent realizations are calculated. For this purpose, all the 45 tags from each realization are measured five times successively. So, the possible number of inter-tag combinations between all different tags within a realization is $C_2^{45\times5}$ - $45C_2^5 = 24750$, and the possible number of intra-tag combinations to compare all repetitive measurements for all tags from one realization is $45C_2^5 = 450$. The signal similarity is calculated by using CS [3.1] and CC_{max} [3.2] for FD and TD signals, respectively, as discussed in sections 3.1 and 5.1.

Figure 5.23 shows the intra- and inter-tag CS and CC_{max} similarity distributions for the FD and TD similarity analyses, respectively, along with their PFP and PFN for the first realization. The procedure for the calculation of the PFP and PFN is similar to the procedure explained in section 3.4. The central tendencies of intra-tag similarity distributions (Figures 5.23(a) and (b)) in the FD and TD analyses are close to unity with minimum values of 0.989 and 0.986, respectively. This behavior validates the repeatability of the individual tags, that is, when the same tag is measured multiple times the result is close to a value of 1. The spreads of inter-tag similarity distributions (Figures 5.23(a) and (b)) are quite wide in comparison to those of intra-tag similarity distributions ranging from 0.142 to 0.993 and from 0.587 to 0.996 for the FD and TD similarity analyses, respectively. It can be observed that the intra- and inter-tag similarity distributions (Figures 5.23(a) and (b)) exhibit overlaps between them, producing PEs of 5.29% and 5.02% in FD and TD analyses, respectively.

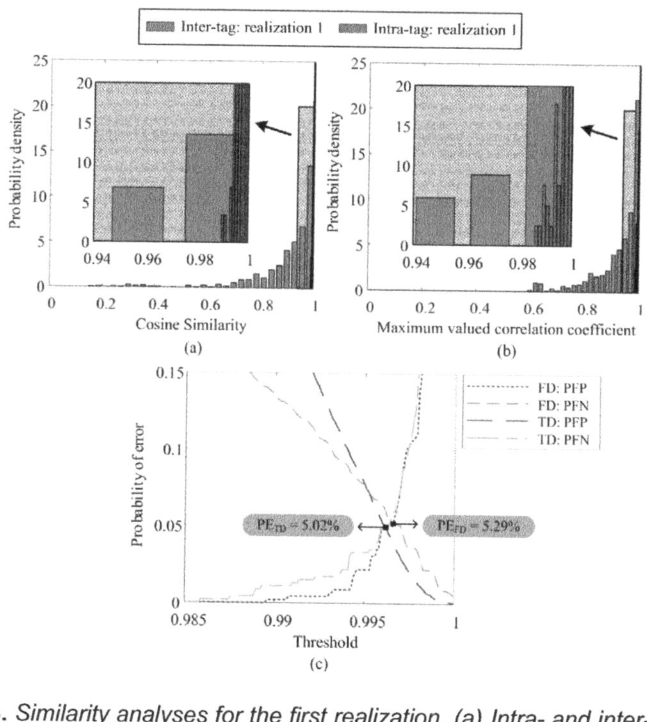

Figure 5.23. *Similarity analyses for the first realization. (a) Intra- and inter-tag cosine similarity distributions for FD analysis. (b) Intra- and inter-tag maximum valued correlation coefficient distributions for TD analysis. (c) Probabilities of false positive and false negative for both FD and TD analyses. For a color version of this figure, see www.iste.co.uk/ali/RFID.zip*

For the authentication results within the second realization, Figure 5.24 shows the intra- and inter-tag CS and CC_{max} similarity distributions for the FD and TD similarity analyses, respectively, along with their error rates (i.e. the PFP and PFN). Intra-tag similarity distributions (Figure 5.24(a) and (b)) in the FD and TD analyses are centered close to unity with minimum values of 0.99 and 0.988, respectively. On the other hand, the inter-tag similarity distributions (Figure 5.24(a) and (b)) exhibit a wide spread ranging from 0.47 to 0.995 and from 0.618 to 0.996 for the FD and TD similarity analyses, respectively. In this case too, the intra- and inter-tag similarity distributions (Figures 5.24(a) and (b)) exhibit an overlap between them, producing PEs of 5.16% and 5.93% in FD and TD analyses, respectively.

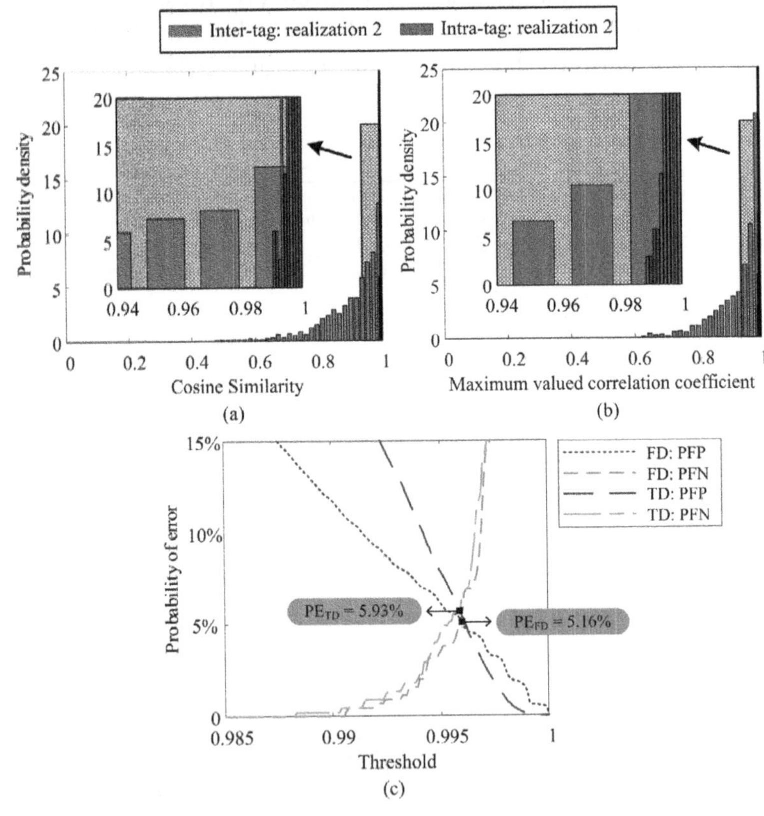

Figure 5.24. *Similarity analyses for the second realization. (a) Intra- and inter-tag cosine similarity distributions for FD analysis. (b) Intra- and inter-tag maximum valued correlation coefficient distributions for TD analysis. (c) Probabilities of false positive and false negative for both FD and TD analyses. For a color version of this figure, see www.iste.co.uk/ali/RFID.zip*

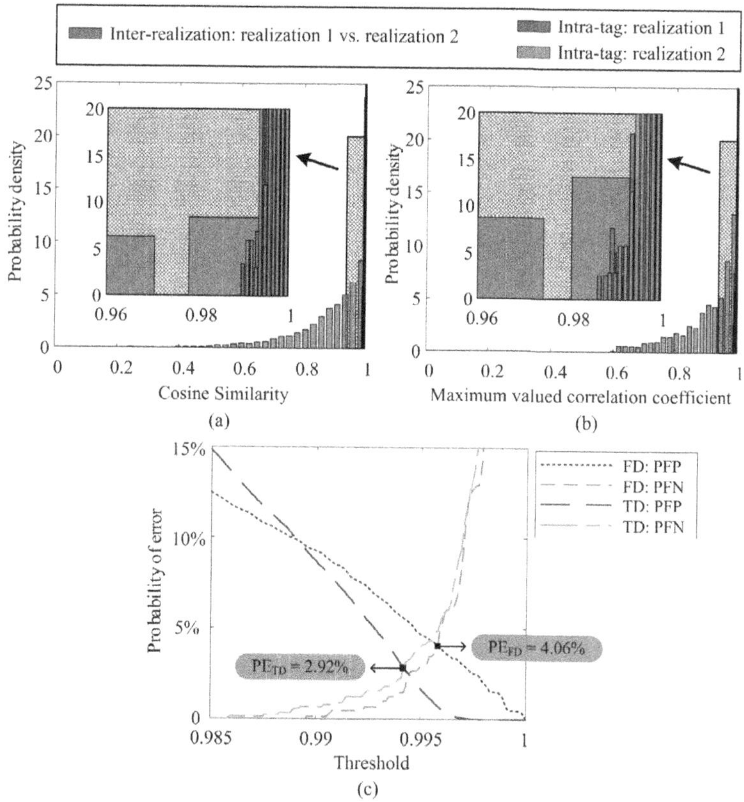

Figure 5.25. *Similarity analyses for the first realization versus second realization. (a) Intra-tag and inter-realization cosine similarity distributions for FD analysis. (b) Intra-tag and inter-realization maximum valued correlation coefficient distributions for TD analysis. (c) Probabilities of false positive and false negative for both FD and TD analyses. For a color version of this figure, see www.iste.co.uk/ali/RFID.zip*

5.4.2. *Authentication across different realizations*

All 45 tags related to each fabrication (see Figure 5.22) are realized at the same time on one piece of substrate. From this fact, the fabrication tolerance would have an almost equal effect on all the 45 structures. So, it is worth calculating the similarity results across different realizations. For a comparison between two independent realizations, the possible number of inter-realization combinations between all different tags is $(45\times5)^2 = 50{,}625$, while the number of intra-tag combinations for each realization is the same as calculated above, that is, $45C_2^5 = 450$. For the comparison between the first realization and second realization, Figure 5.25 shows the intra-tag and inter-realization CS and CC_{max} similarity distributions for the FD and TD similarity analyses, respectively, along with their error rates (i.e. PFP and PFN). The procedure for the calculation of the PFP and PFN is similar to the procedure explained in section 3.4. In this comparison, the spreads of inter-realization similarity distributions (Figure 5.25(a) and (b)) are also quite wide in comparison to intra-tag similarity distributions ranging from 0.23 to 0.993 and from 0.588 to 0.994 for the FD and TD similarity analyses, respectively. The overlaps among the intra- and inter-realization similarity distributions (Figures 5.25(a) and (b)) produce PEs of 4.06% and 2.92% in FD and TD analyses, respectively. Such values of PE are quite low and comparable to the average PE for the fingerprint verification system ranging from 2.07% to 4.03% found in (Cappelli *et al.* 2006) and the best average PE of 0.05% among different fingerprint evaluation campaigns found in (Maltoni *et al.* 2009, Chap. 4).

5.4.3. *Characterization of the natural randomness*

The existence of natural dimensional variations in the realized prototypes of two intermittent realizations is analyzed by microscopic analysis. A Mighty Scope 5M digital microscope by Aven tools, as shown in Figure 5.18, is used. The calibration of the system is done by the USAF 1951 resolution test chart. The resolution of the system is approximately 3 μm per pixel. First, the digital images of chipless tags are captured and then Hough transform is applied to calculate geometrical dimensions in the microscope images. A sample image of all measured dimensional parameters using the Hough transform is shown in Figure 5.26. In the inset, the Hough lines along the arms of the first C-folded scatterer of the tag N1 are shown.

It is clear from [3.5] and [5.2] that the L' and g are associated with the f_r and Q, respectively. For this reason, eight L_m and four g_m parameters for all the 45 tags of both intermittent realizations are measured (see Figure 5.26). In theory (see [3.5] and [5.2]), only one value for each L' (or alternatively L) and g parameters is expected. However, in practice, because of quad coupled scatterers, the parameters' values can change from one element to another (see Figure 5.26) that can contribute to the randomness of the response. Figure 5.27(a) shows a comparison of the measured arms' lengths L_m for both intermittent realizations along with the value of the theoretical arms' length L, and Figure 5.27(b) shows a comparison of the measured spacings between the arms g_m for both intermittent realizations along with the theoretical value of the spacing between the arms g. It can be observed that the mean values of the distributions of both measured parameters (L_m and g_m) for two realizations are clearly separated. The mean values of the L_m for the realizations 1 and 2 are $L_{m(\mathrm{R1})} = 3.13$ mm and $L_{m(\mathrm{R2})} = 3.16$ mm, respectively. The mean values of the g_m for the realizations 1 and 2 are $g_{m(\mathrm{R1})} = 2.50$ mm and $g_{m(\mathrm{R2})} = 2.49$ mm, respectively.

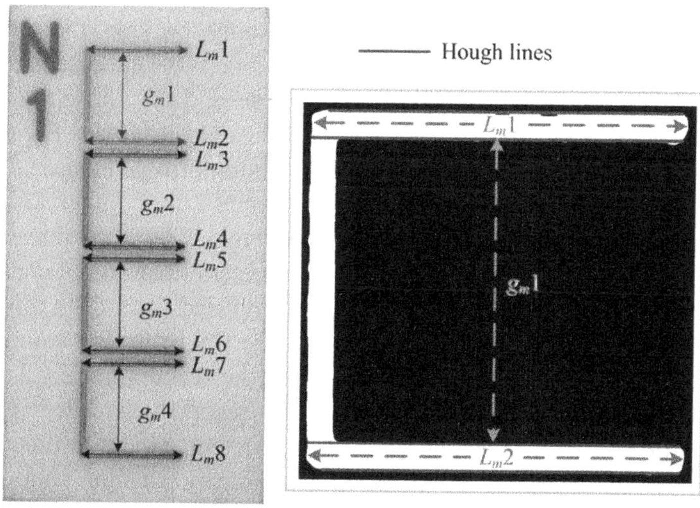

Figure 5.26. *Measurement dimensional parameters from microscope digital images using the Hough transform. In the inset, an example of a C-folded scatterer along with its dimensional parameters is shown. For a color version of this figure, see www.iste.co.uk/ali/RFID.zip*

To associate the natural random dimensional variations with the signals' similarity variations (merely due to aspect-independent parameters), the combinations of two tags from two distributions (see Figure 5.27) exhibiting maximum and minimum average dimensional variations (i.e. extreme variations) along L_m, ΔL_m, and along g_m, Δg_m are detected. In this way, we have four extreme cases. For each case from four extreme cases, f_r and Q of both selected tags from two realizations are extracted from their signals (average of five repetitive measurements) and then the variation of f_r, Δf_r, and the variation of Q, ΔQ, between these two tags are calculated. Table 5.1

illustrates the detected tags from two realizations for maximum and minimum average variations ΔL_m and Δg_m parameters along with their extracted Δf_r and ΔQ. For simplicity, we assume that Δf_r is merely related to ΔL_m (see [3.5]) and ΔQ is merely related to Δg_m (see [5.2]). For this reason, for extreme cases of ΔL_m, the values of ΔQ are neglected (highlighted in red) and the values of Δf_r are taken (highlighted in green). Similarly, for extreme cases of Δg_m, the values of Δf_r are neglected (highlighted in red) and the values of ΔQ are taken (highlighted in green). However, this analysis along with our simplified assumption may only provide the first-order approximation of the results. Next, the CS map of quad-scatterer tags presented in Figure 5.14 is taken as a reference and the chosen values are superimposed (in the form of lines) on the CS map to create Figure 5.28. Figure 5.28 describes the range of the similarity change that can happen due to the practical natural randomness in comparison to the theoretical model for these quad-scatterer tags. The theoretical model accounts for only the aspect-independent parameters (that do not contain any measurement or systematic noise). The superimposed lines form a rectangle enclosing an area of similarity variations on the CS map. The values of CS on the CS map beneath these four vertices of the rectangle are $CS_A = 0.99$, $CS_B = 0.99$, $CS_C = 0.77$ and $CS_D = 0.78$, respectively. It is found that the current natural variations are sufficient to be employed for authentication applications by producing similarity change ranging from $CS = 0.99$ to $CS = 0.77$. It is important to note that these variations are purely based on variations of the aspect-independent parameters: Δf_r and ΔQ. The maximum change in CS due to the natural randomness $\Delta CS_{natural}^{max} = CS_A - CS_C = 0.22$ (Figure 5.28) is the same order of magnitude as the simulated results $\Delta CS_4^{CST} = 0.22$ and $\Delta CC_{max4}^{CST} = 0.23$ (Figure 5.8). On the other hand, $\Delta CS_{natural}^{max}$ is less than the change in CS due to purposely applied variation $\Delta CS_4 = 0.6$ (Figure 5.14). The reason for this decrease is that in natural randomness the maximum detected change of ΔL_m is $\Delta L_m^{max} = 57.5\ \mu m$ (see Table 5.1), while the purposely applied variation is 100 μm in Figure 5.14.

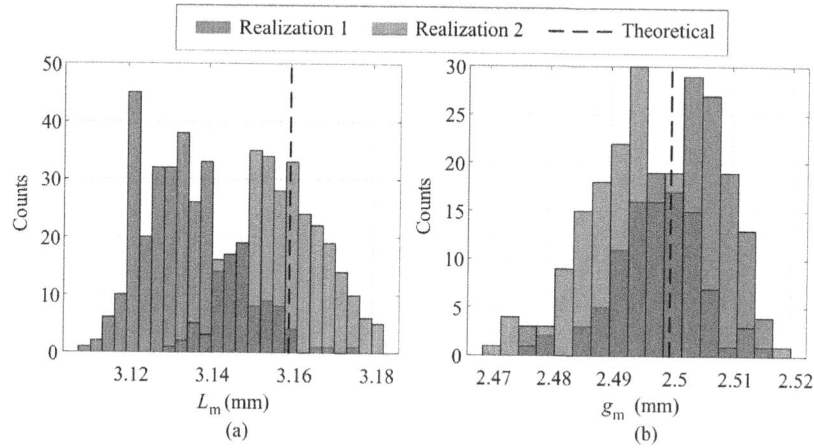

Figure 5.27. *Microscopic dimensional characterization for both intermittent realizations. (a) Measured arms' lengths L_m in comparison to the theoretical value of arms' length L. (b) Measured spacings between the arms g_m in comparison to the theoretical value of the spacing between the arms g. For a color version of this figure, see www.iste.co.uk/ali/RFID.zip*

Distributions	Minimum case				Maximum case			
	Tags	**ΔL_m (μm)**	**Δf_r (MHz)**	**ΔQ**	**Tags**	**ΔL_m (μm)**	**Δf_r (MHz)**	**ΔQ**
$L_{m(R1)}$ vs. $L_{m(R2)}$	$N3_{R1}$ and $N3_{R2}$	≈ 0	13.5	1.24	$N14_{R1}$ and $N28_{R2}$	57.5	94	12.51
		Δg_m (μm)				Δg_m (μm)		
$g_{m(R1)}$ vs. $g_{m(R2)}$	$N2_{R1}$ and $N15_{R2}$	≈ 0	0.5	2.43	$N14_{R1}$ and $N28_{R2}$	37.4	94	12.51

Table 5.1. *Four extreme cases of dimensional variations of average L_m and g_m between realizations 1 and 2. For a color version of this table, see www.iste.co.uk/ali/RFID.zip*

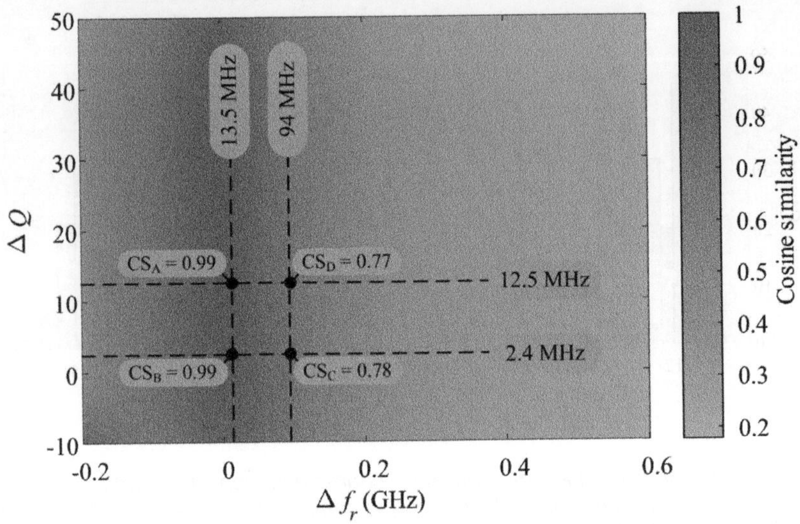

Figure 5.28. *Range of the similarity change happened due to the natural random variations for quad-scatterer C-folded tags. For a color version of this figure, see www.iste.co.uk/ali/RFID.zip*

We can argue that the detected Δf_r and ΔQ parameters may be due to the variation of ε_r of the employed substrate (i.e. Rogers RO4003) from one sample to another. For this reason, we conducted simulations for quad-scatterer tags using its nominal dimensional parameters (see Figure 5.8) while changing ε_r from 3.38 to 3.38 ± 0.05 that is provided by the manufacturer. From simulated results, we found that an average Δf_r due to changes in ε_r is $\Delta f_r^{\Delta\varepsilon_r} \approx 62$ MHz. This value of $\Delta f_r^{\Delta\varepsilon_r}$ is less than the detected natural $\Delta f_r^{\max} = 94$ MHz (see Table 5.1), which signifies that the detected natural randomness is not merely due to the variation of ε_r. As an additional benefit, the variation of ε_r would contribute to the natural randomness to elevate the potential of the chipless RFID technology for authentication applications.

5.4.4. Generalization of the proposed method

From section 5.3.2, the worst inter-realization PE equals 4.06% (see Figure 5.25(c)) in the CS case. Such a value of PE may be large for the practical implementation of the proposed method. It is important to note that the natural random variations along the geometrical dimensions of the two tags belonging to two different realizations are independent. Therefore, this independent nature of the random variations can be exploited to decrease the PE by increasing the number of employed scatterers in the design of chipless RFID tags, as shown in Figure 5.29(a). For this purpose, the arms' length of the further additive quad-scatterers can be decreased by a step value dL to shift their f_r to larger values. With a value $dL = 100\ \mu$m, it can be observed from Figure 5.29(a) that the peak apex related to each scatterer is well separated. This procedure would aid in the reading of all the scatterers in the design of chipless RFID tags simultaneously. As the realization error for the two scatterers is independent, we can use the Bayes rule:

$$p(a|b) = p(a). \qquad [5.3]$$

Then,

$$p(a,b) = p(a)\,p(b). \qquad [5.4]$$

For N number of scatterers using [5.4], the worst inter-realization PE = 4.06% (see Figure 5.25(c)) in CS case can be written as:

$$\text{PE} = (0.04)^n, \qquad n = 1, 2, \ldots N \qquad [5.5]$$

where n is the number of scatterers to be employed in the design of the tags developed for the authentication process. Figure 5.29(b) shows an exponential decay of PE with the increase in the number of scatterers using [5.5]. To achieve a PE = 0.05% found in (Maltoni *et al.* 2009, Chap. 4) from the proposed chipless RFID approach, three quad-scatterers are required in the design as mentioned in Figure 5.29(b).

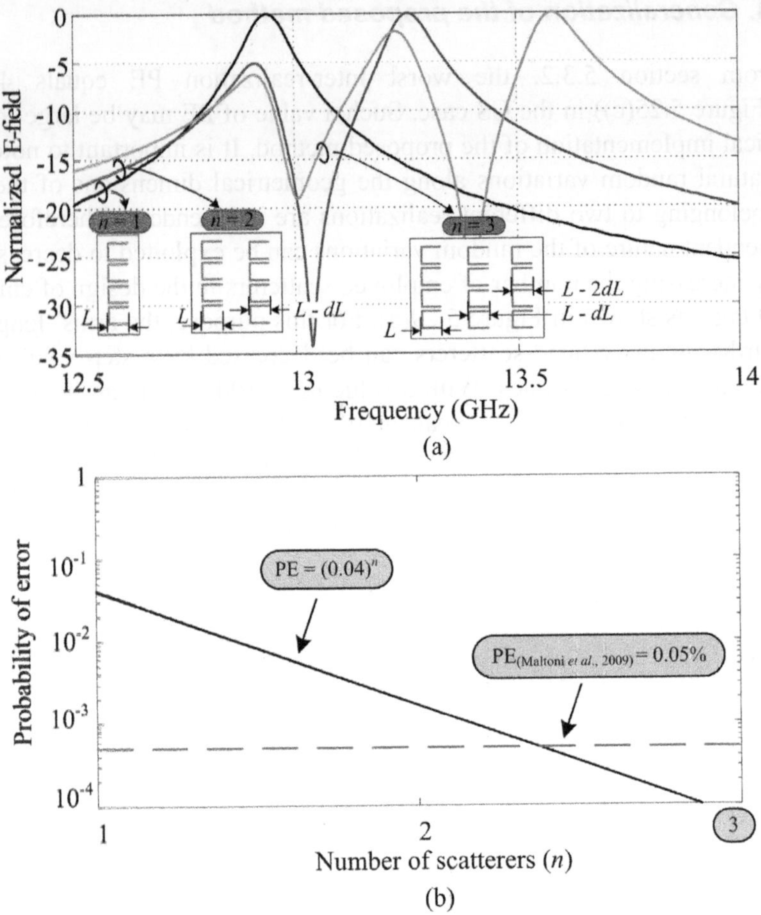

Figure 5.29. *Exponential decay of the probability of error with the increase in the number of employed scatterers in the design of tags. (a) Simulated responses of the tags while increasing the number of employed scatterers in the design of tags. (b) Exponential decay of the probability of error with an increase in the number of employed scatterers in the design of tags*

5.4.5. *Final remarks on the constraints*

We can argue that the cost for the examination of the authenticity in the proposed method is high because of the cost of the VNA. Such a high cost is in the same range of the other available highly secure authentication solutions (e.g. biological authentication solutions or X-ray-based authentication solutions), where the cost of the equipment to examine the authenticity is also high. On the other hand, this high-cost equipment is not intended for the common consumer to verify the authenticity of to-be-purchased items. Even this high-cost equipment may not be intended for the suppliers of the goods, the distributors, the food and drug administration, or the representatives of pharmaceutical companies. Instead, this is for customs inspectors, law enforcement organizations and the court of justice, as these highly secure solutions are specifically for high-end luxury items.

The cost of the system can be reduced by employing the commercially available chipless RFID readers (Garbati *et al.* 2018, Chap. 4: "RFIDTECH large reader"). The IR-UWB reader operates in the frequency band ranging from 3.1 GHz to 10.6 GHz, as discussed in Garbati *et al.* (2018, Chap. 4). The cost of this reader is comparable with the classical RFID reader used in logistics and supply chain management. To use this IR-UWB chipless reader (Garbati *et al.* 2018, Chap. 4), the action of scaling the design of the tags to a lower frequency of operation within UWB poses a slight enlargement in the size of the tags. As another alternative, RFIDTECH large reader can also be used. The band of operation of this chipless reader ranges from 22 GHz to 26.5 GHz. In this case, the size of the tags will be even smaller than the size presented in this chapter because of the action of scaling the design of the tags to a larger frequency of operation, which is advantageous.

5.5. Conclusion

First, we have shown that the four-coupled C-folded scatterer-based chipless tag is a better choice than the single C-folded scatterer-based chipless tag due to its high Q.

To show the minimum detectable dimensional variation, we have fabricated three groups of tags (quad C-folded scatterer tags) exhibiting distinct arms' length which mimic the process variations as in the fabrication process.

We have used a second-order bandpass filter model to describe the antenna mode of a C-folded resonator. Owing to this proposed model along with the aspect-independent parameters, we have demonstrated that two groups of the tags exhibiting a measured mean variation of approximately 25 μm along the arms' length can be discriminated with a success rate greater than 99.9% even in a realistic outdoor environment. A good agreement was shown between the theoretical model and the experimental results. According to the validated model, the proposed discrimination technique could potentially differentiate two close structures whose dimensions would vary in the accuracy range of the classical manufacturing technique used to realize cheap tags.

Finally, natural dimensional variations in the design of C-folded tags were analyzed to be used for the authentication applications. For this purpose, quad-scatterer tags were chosen because of their sharp slope of dissimilarity. Chipless tags were realized two times intermittently, where each realization was constituted of 45 tags. The two different realizations share the same company, same PCB technology, but a different film mask to ensure the natural dimensional randomness. The similarity analyses were conducted inside each realization and also between two intermittent realizations. The achieved probability of error was comparable to the various fingerprint evaluation campaigns found in the literature. The existence of the natural dimensional variations was also confirmed by the microscopic dimensional analysis for two intermittent realizations. Then, the minimum and maximum changes between the intermittent realizations were also linked to the similarity change to show an intuitive range. It was found that the existing natural variations were enough to produce similarity change to be employed for authentication applications. Finally, the technique was also generalized to reduce the probability of error to a significant level.

For greater security, the chipless tags can also be inserted into the pulp of the packaging of the product or in the layers of packaging during the corrugation process of the packaging.

6

Chipless Authentication Using Inkjet-Printed PET Tags

6.1. Introduction

In this chapter, a chipless authentication method using natural randomness inherent in the inkjet printing process is presented. The chipless RFID tags printed with a cheap off-the-shelf inkjet printer with metallic ink are utilized. The design of the chipless tags is explicitly optimized for inkjet printing based on the resolution of the inkjet printer. These optimized chipless tags are very difficult to duplicate, as their unique backscattered EM responses depend on the proximate coupling among the vertex-to-vertex adjacent square geometries, which occurred naturally due to randomness in the inkjet printing. The performance of the system has been analyzed by the highly accurate VNA-based reader as well as by a low-cost IR-UWB reader. The achieved PE is comparable to the various fingerprint evaluation campaigns found in the literature. This chapter is organized as follows:

– section 6.2 is dedicated to the optimization of the design of a conventional C-folded scatterer to elevate its sensitivity to the inkjet printing;

– sections 6.3 and 6.4 provide the validation of the proposal by presenting the performance of the optimized chipless tag (with natural randomness) using a highly accurate VNA-based chipless reader and using a low-cost impulse IR-UWB chipless reader, respectively;

– section 6.5 presents the conclusion.

As discussed in Chapter 3, the similarity comparisons are performed in the FD using CS [3.1] and in the TD [3.2] using CC_{max} (see Ali *et al.* 2018, 2019, 2020).

6.2. Optimization of chipless tags to exploit natural randomness inherent in inkjet printing

From the discussion in Chapter 3 (section 3.6), it seems that it might not be possible to exploit process randomness inherent in the inkjet printing technology to create unique EM signals from the conventional design of the chipless tags. For this purpose, we adopted an optimization of the design of the chipless tags to elevate their sensitivity to the randomness inherent in inkjet printing. Figure 6.1 shows a zoom photograph of one of our realizations. It can be seen that from a vertex-to-vertex separated digital design (i.e. supplied to the inkjet printer), a randomly connected or disconnected realization can be printed. It is important to note that these random variations are happening during the fabrication process (inkjet printing) but not from the ideal digital design. Therefore, simple vertex-to-vertex adjacent metallic geometries can behave like electrical switches, as depicted in Figure 6.1. A vertex-to-vertex connection between two metallic geometries is analogous to an electrical switch in a closed state, while a vertex-to-vertex separation between two metallic geometries is similar to an electrical switch in an open state. Such a vertex-to-vertex adjacent design can be realized from a simple square check pattern to prove the concept. Next, the square check patterns will be incorporated to optimize the design of chipless RFID tags to be used for authentication.

As discussed in Chapters 3 and 5, the response of a C-folded scatterer is mainly driven by two key dimensional parameters (see [3.5] and Figure 6.2(a)): the spacing between the two arms g and the length L (or alternatively L'). Such a dependence of the response on the geometrical dimensions is very suitable for the use of the square check pattern to optimize the design of the classical C-folded scatterer for authentication application. Figure 6.2 provides the concept of the optimization from the design of conventional C-folded scatterer (Figure 6.2(a)) to the sensitive C-folded scatterers (Figure 6.2(b) and (c)). For a C-folded scatterer, the RCS is directly associated with the gap g between the arms, as discussed in Vena *et al.* (2011), with comparable dimension, as shown in Figure 6.2. For this reason, the g is increased to a value of 3.5 mm in the design of the optimized tags compared to the conventional design discussed in Chapter 3. To demonstrate the optimization, the length L of the conventional C-folded scatterer is divided into two parts: the primary length L_1 and the secondary length L_2. Figure 6.2(b) and (c) shows the layouts of the optimized design of the C-folded scatterer exhibiting check patterns with three-square metallic elements and with five-square metallic elements along the secondary length of the arms L_2, respectively.

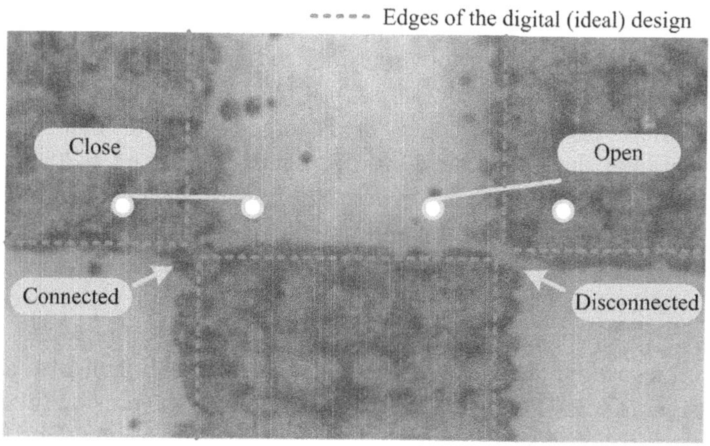

Figure 6.1. *An analogy of the vertex-to-vertex adjacent metallic geometries to electrical switches, where the vertex-to-vertex connected geometries behave like an electrical switch in a closed state, while the separated geometries behave like an electrical switch in an open state. For a color version of this figure, see www.iste.co.uk/ali/RFID.zip*

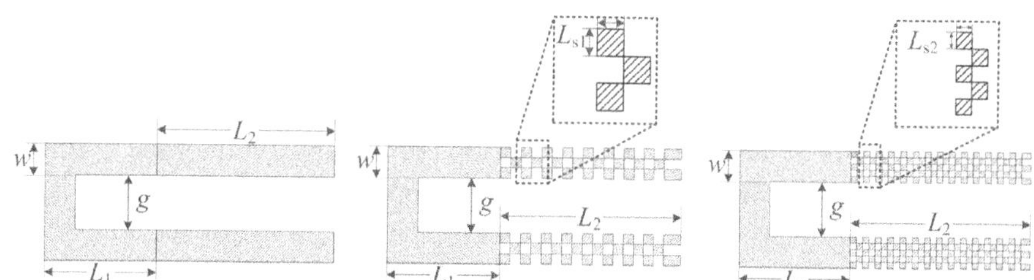

Figure 6.2. *Layouts of the C-folded scatterers. (a) Conventional design of the C-folded scatterer. (b) Optimized design of the C-folded scatterer exhibiting a check pattern with three-square metallic elements. (c) Optimized design of the C-folded scatterer exhibiting a check pattern with five-square metallic elements. The L_1 = 5.35 mm, L_2 = 17.15 mm, w = 2 mm, g = 3.5 mm, L_{s1} = 0.67 mm and L_{s2} = 0.4 mm. For a color version of this figure, see www.iste.co.uk/ali/RFID.zip*

The substrate used in this chapter is a PET printable film (Novele™ IJ-220) with thickness $t = 140$ µm. The electrical characterization of this PET printable film is conducted using the cavity resonator from Damaskos, Inc., which yields the relative permittivity $\varepsilon_r = 2.99$ and dielectric loss tangent $\tan\delta = 0.015$. Such a PET-based substrate is chosen because of its low cost, low loss and directly printable properties. The simulated backscattered FD responses of the layouts of C-folded scatterers presented in Figure 6.2 are shown in Figure 6.3. The layouts are illuminated by a plane wave using a full-wave EM simulator (CST MWS), as shown in the inset of Figure 6.3. From [3.5], it is obvious that the f_r primarily depends on the length of the arms L (or alternatively L') of the C-folded scatterer. Similarly, the simulation results presented in Figure 6.3 show that the f_r for the conventional design of the C-folded scatterer (Figure 6.2(a)) changes from 3.31 GHz to 7.79 GHz as the length of arms changes from $L_1 + L_2$ to L_1. The simulation results for the optimized chipless tags exhibiting the total length of arms $L_1 + L_2$ present the values of f_r close to the value of f_r of the conventional chipless tag (see the blue dashed line and violet two-dashed line). However, in practice, f_r of the optimized chipless tags would produce random values within the frequency band of the variation highlighted as a gray region in Figure 6.3. These variations would be due to the possibility of having random electrical conductivity among the adjacent square geometries depending on the random position of the drops in inkjet printing, as depicted in Figure 6.1. It would ultimately produce unique patterns to be appropriate for authentication. These unique tags are very difficult to duplicate, as the backscattered EM responses depend on the proximate coupling among the possible separated squares, which occurs due to randomness inherent in inkjet printing.

In practice, the optimized scatterers (Figure 6.2(b) and (c)) would be sensitive to each existing square geometry in the check patterns. To demonstrate the sensitivity of the optimized scatterers, we purposely varied the length of the beginning interior and exterior square elements of the lower arm of the C-folded scatterer. Figure 6.4 illustrates the purposely applied variations of the lengths of the beginning interior and exterior square elements of the lower arm of the C-folded scatterer by displacements d_{in} and d_{out}. The d_{in} and d_{out} correspond to the displacement of the length of the interior and exterior square elements, respectively. The superscript in displacement parameters (d_{in} and d_{out}) shows the number of square elements in the check patterns: 3 or 5. For both optimized scatterers (Figure 6.4), while changing the d_{in} from −100 µm to 100 µm with a step of 50 µm, the d_{out} is kept fixed at 0 µm (i.e. the connected position for the exterior square element) and vice versa.

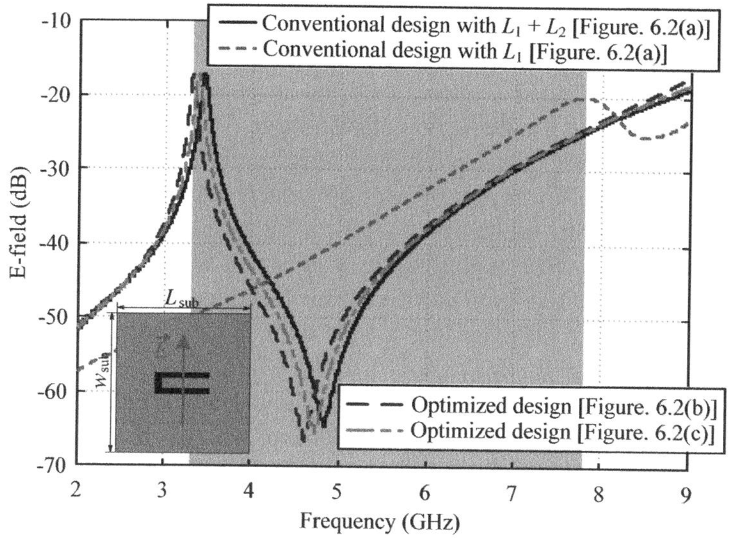

Figure 6.3. *Simulated backscattered responses of the layouts shown in Figure 6.2. The region highlighted in gray shows the frequency band of the possible variation of f_r due to randomness. The substrate's dimensions are L_{sub} = w_{sub} = 48 mm. For a color version of this figure, see www.iste.co.uk/ali/RFID.zip*

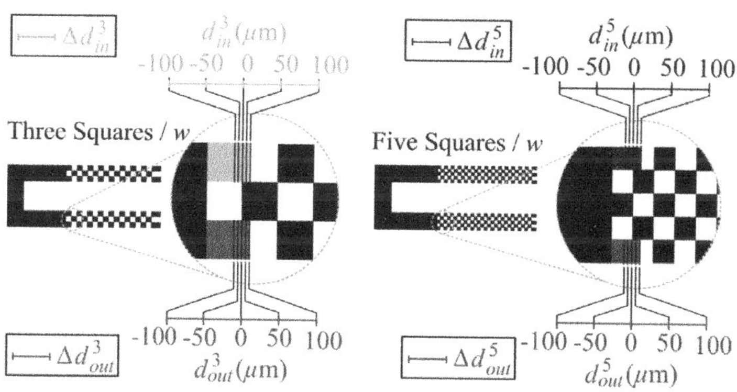

Figure 6.4. *Illustrations of variations in the lengths of the beginning interior and exterior square elements of the lower arm of the C-folded scatterer to check the sensitivity of optimized scatterers. (a) Three-square check-patterned C-folded scatterer (Figure 6.2(b)). (b) Five-square check-patterned C-folded scatterer (Figure 6.2(c)). The d_{in} and d_{out} correspond to the displacement of the length of the interior and exterior square elements, respectively. The superscript shows the number of square elements of the check pattern. For a color version of this figure, see www.iste.co.uk/ali/RFID.zip*

For all the purposely applied variations in both optimized scatterers, the simulations are conducted on the PET substrate in the same way as discussed for the classical C-folded scatterer (see the inset of Figure 6.3). Table 6.1 outlines the change in the frequency of resonance Δf_r for each applied change in displacements Δd to the beginning interior and exterior square elements (Δd_{in} or Δd_{out}) of the lower arm of both optimized C-folded scatterers (Figure 6.2(b) and (c)). It can be observed that for each subjected square element, the Δf_r is maximum (see highlights in Table 6.1) while Δd (Δd_{in} or Δd_{out}) is ranging from −50 µm to 0 µm (i.e. from disconnection to a connection). A C-folded scatterer behaves like a slot with a high concentration of the current density in the interior region than the exterior region of the two arms (see Rance *et al.* 2016, Figures 3 and 4). Hence, the optimized designs are more sensitive to the interior square element than to the exterior square element (i.e. Δf_r for the interior square elements is larger than Δf_r for the exterior square elements when Δd (Δd_{in} or Δd_{out}) ranges from −50 µm to 0 µm). From Table 6.1, it can also be observed that apart from maximum effective Δd (i.e. −50 µm to 0 µm and with which Δf_r is larger), the optimized designs of the C-folded scatterer are also sensitive to the other displacements to a minor level. It is important to note that the simulated sensitivity analysis by changing the length of specific individual square elements is presented (Figure 6.4 and Table 6.1) just to show the idea of the sensitivity of the designs merely for one square element at an instant. In fact, these random changes would happen naturally for all the square elements of the patterns with a sort of cumulative effect during the fabrication process. The absence or presence of a connection will be directly done by the printer itself due to its low resolution.

For the realization of the square check-patterned C-folded scatterer chipless tags, we used the Epson C88+ printer with a JS-B25P silver conductive ink without the annealing process. Prior to the printing of the tags, the minimum resolution and randomness of ink particles have been characterized, as shown in Figure 6.5, where the randomness of the ink droplets can be observed. The digital design (mask) of the solid circular shapes of known sizes is supplied to the printer using a PET printable film as printing media. Then, the dimensions of the inkjet-printed output are analyzed using a microscope. In inkjet printing, the output usually consists of two types of ink droplets: main droplets and satellite droplets (Carnahan and Hou 1977; Hoath *et al.* 2012). Hence, if we consider the smaller and separated ink droplets as satellite droplets to ignore them, the minimum resolution of the employed system can be calculated around values ranging from 55 μm to 116 μm as a first approximation.

Change in displacement parameters (Δd) (μm)		Three-square check pattern	Five-square check pattern
		Δf_r (MHz)	Δf_r (MHz)
Δd_{in}	−100 to −50	0.1	0.1
	−50–0	35.4	10.8
	0–50	0.6	0.3
	50–100	1.2	0.6
Δd_{out}	−100 to −50	0.6	0.1
	−50–0	22.8	5.7
	0–50	0.3	0.1
	50–100	0.9	0.3

Table 6.1. *Sensitivity (change in the frequency of resonance Δf_r) for each applied change in displacements Δd to the beginning interior and exterior square elements (Δd_{in} or Δd_{out}) of the lower arm of both optimized C-folded scatterers (Figure 6.2(b) and (c))*

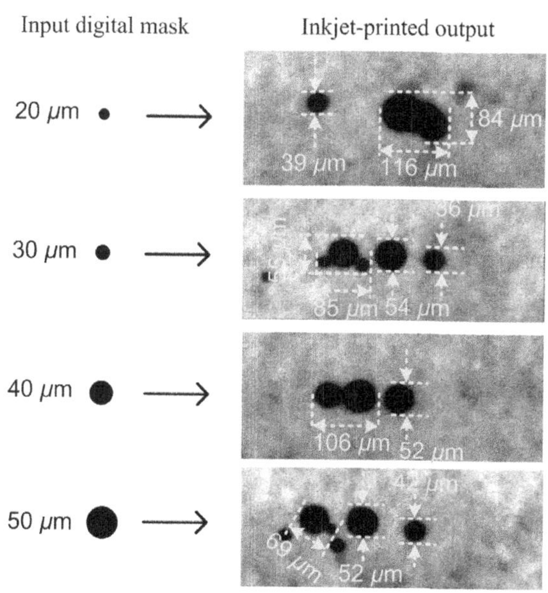

Figure 6.5. *Randomness in inkjet printing using the Epson C88+ printer with a JS-B25P silver conductive ink on a PET printable film. For a color version of this figure, see www.iste.co.uk/ali/RFID.zip*

Regarding the inkjet printing of the digital mask of the check patterns exhibiting vertex-to-vertex connected squares (Figure 6.2(b) and (c)), we have testified that the inkjet-printed output always produces a well-connected check pattern, which is reducing the opportunity of exploiting the randomness of inkjet printing for authentication purposes. For this reason, we designed the input digital mask of check patterns with a vertex-to-vertex gap d_s of 70.7 µm (i.e. hypotenuse distance d_s of a 50 µm^2 square geometry) among the square elements, as shown in Figure 6.6. This value of the d_s is chosen in the middle of the inkjet printer resolution range (i.e. from 55 µm to 116 µm) to exploit the randomness inherent in inkjet printing efficiently.

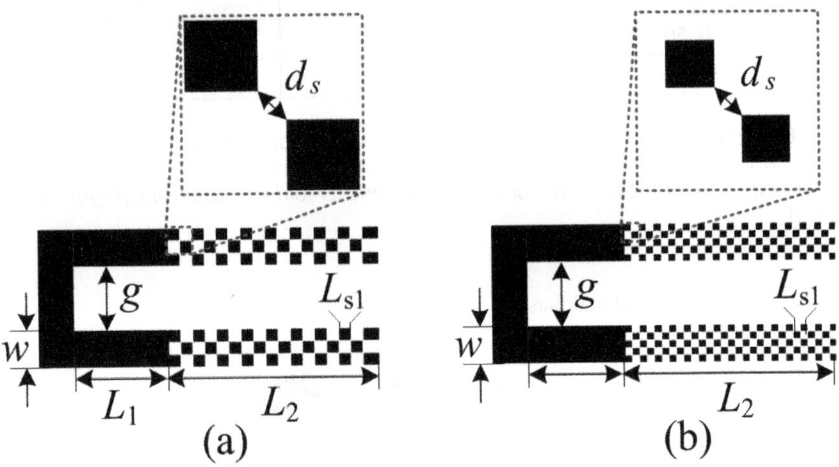

Figure 6.6. *The input digital mask of check patterns with a vertex-to-vertex gap d_s of 70.7 µm (i.e. hypotenuse distance d_s of a 50 µm^2 square geometry) among the square elements. (a) Input digital mask of three-square check-patterned C-folded scatterer. (b) Input digital mask of five-square check-patterned C-folded scatterer. The L_1 = 5.35 mm, L_2 = 17.15 mm, w = 2 mm, g = 3.5 mm, L_{s1} = 0.63 mm, L_{s2} = 0.36 mm and d_s = 70.7 µm. For a color version of this figure, see www.iste.co.uk/ali/RFID.zip*

Thirty prototypes of identical tags for each digital mask (Figure 6.6) are printed on a PET printable film. Figure 6.7 shows the first tag based on three squares (Figure 6.7(a)) and the first tag based on five squares (Figure 6.7(b)) check-patterned C-folded scatterer.

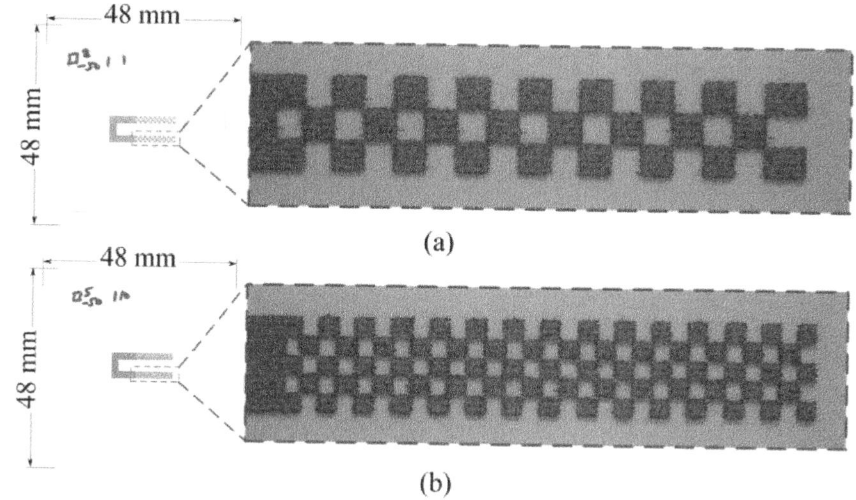

Figure 6.7. *Top view of inkjet-printed tags based on the check-patterned C-folded scatterer on a PET printable film. (a) The first tag based on three-square check-patterned C-folded scatterer. (b) The first tag based on five-square check-patterned C-folded scatterer. For a color version of this figure, see www.iste.co.uk/ali/RFID.zip*

6.3. Authentication using VNA-based chipless reader

For an accurate characterization, free space measurements using a VNA (Agilent 5222A) are performed. Figure 6.8 shows the measurement setup in an anechoic environment with a Satimo (QH2000) quad ridged open boundary antenna (2–32 GHz). The source power of VNA is equal to –5 dBm. The frequency sweep ranging from 3 to 8 GHz with 10001 points is used. The reason for this chosen spectrum is that the frequency band of the variation of the optimized chipless tags resides in this range (see the gray highlighted region in Figure 6.3). The tags are placed at a distance r =16.5 cm from the antenna. The separation distance between two antennas is e = 2.7 cm. Even though the measurements are done in a bistatic co-polarization configuration with two antennas, we present the results based only on the reflection coefficient S_{11} in this chapter for the VNA-based chipless reader. The results based on S_{11} appear to be more promising for practical situations due to the use of only one antenna (fewer resources).

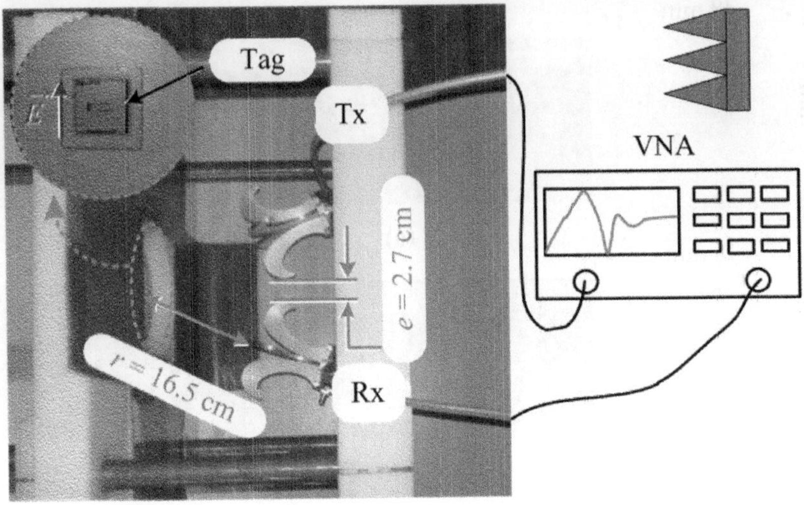

Figure 6.8. *Measurement setup in an anechoic environment with the VNA-based chipless reader. For a color version of this figure, see www.iste.co.uk/ali/RFID.zip*

The reflection coefficient S_{11} of the PET inkjet-printed tags (Figure 6.7) is measured. Each tag is measured five times successively, by removing and repositioning the tag at its position during each measurement trial to check the repeatability. The rest of the post-treatment on the measured signals (i.e. background normalization and time windowing) is the same as discussed in sections 3.5.1 and 4.3 (Ali *et al.* 2018, 2019, 2020). For the time windowing, the early part of the TD signals up to 2.2 ns is discarded and the subsequent part of 10 ns is extracted, as shown in Figure 6.9. The effective length windowed TD signal is of 100 points by neglecting the padded zeros. The padded zeros do not contribute to the similarity while comparing the signals. The windowed FD signal exhibits 9,800 points.

The first measurement of S_{11} in the form of windowed FD responses for all the 30 C-folded scatterer tags patterned with three squares (Figure 6.7(a)) and with five squares (Figure 6.7(b)) is shown in Figure 6.10(a) and (b), respectively.

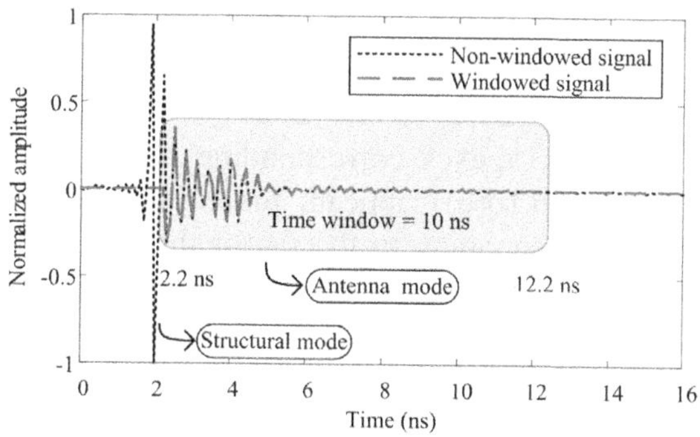

Figure 6.9. *Time windowing to discard the structural mode and to extract the antenna mode. The start time of window is 2.2 ns and the duration of the time window is 10 ns. For a color version of this figure, see www.iste.co.uk/ali/RFID.zip*

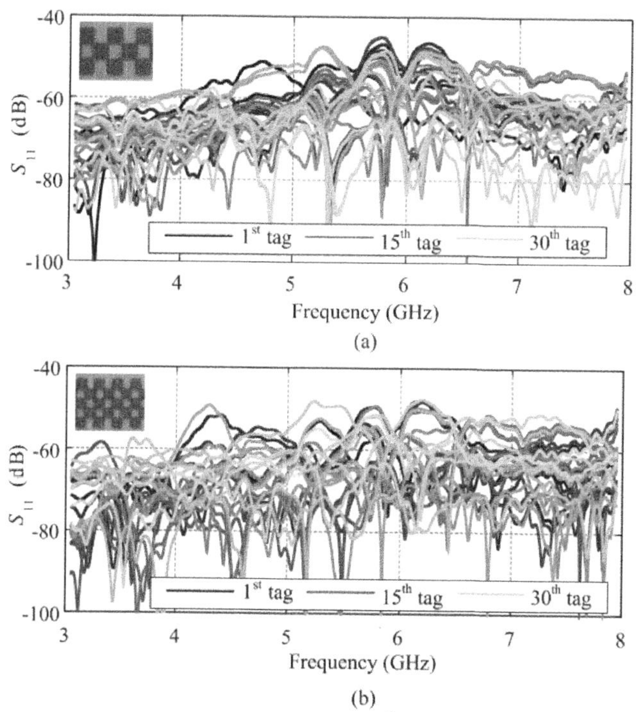

Figure 6.10. *First measurement of S_{11} in the form of windowed FD responses for all the 30 tags. (a) For three-square check-patterned C-folded scatterer tags. (b) For five-square check-patterned C-folded scatterer tags. The corresponding zoom photographs of the inkjet-printed patterns are shown in the insets. For a color version of this figure, see www.iste.co.uk/ali/RFID.zip*

The randomness that happened during the inkjet printing process can be observed from the randomness of the position of the peak apexes corresponding to the frequencies of resonance (f_rs) of the tags. Furthermore, the position of the peak apexes corresponding to the f_rs for five-square check patterns (Figure 6.10(b)) appears more random than the position of the peak apexes corresponding to the f_rs for three-square check patterns (Figure 6.10(a)). This larger randomness of the peak apexes for five-square check patterns is due to the larger surface area exposed to the inkjet printing that in turn gives rise to larger variations of proximate couplings in the design. On the other hand, Figure 6.11 shows an example of the repeatability of the measurement system by presenting the repetitive measurements of S_{11} in the form of windowed FD responses for the first sample from the three-square check-patterned C-folded scatterer tag using the VNA-based chipless reader. In the inset, a zoom graph is presented to show the agreement of the repetitive measurements. The perfect repeatability of the measurements can be observed in Figure 6.11.

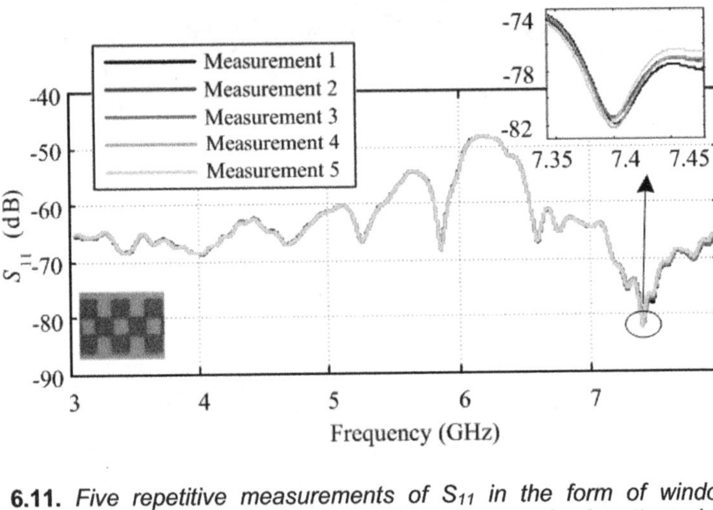

Figure 6.11. *Five repetitive measurements of S_{11} in the form of windowed FD responses for the first sample from the three-square check-patterned C-folded scatterer tags using the VNA-based chipless reader. In the inset, a zoom graph is presented to show the agreement of the repetitive measurements. For a color version of this figure, see www.iste.co.uk/ali/RFID.zip*

For each pattern, after the post-treatment, all the measured signals (i.e. 30 identical tags × 5 measurements for each tag = 150 signals) are compared among them to calculate the similarity level. For all the 30 identical patterned tags for each pattern, the intra-tag combinations can be calculated as $30C_2^5 = 300$, and the inter-tag combinations can be calculated as $C_2^{30\times5} - 30C_2^5 = 24750$. Subsequently, statistical analyses are performed for both FD and TD signals, as explained in Figure 3.3. For the rest of this chapter for each case, the FD analysis (intra- and inter-tag CS distributions) is performed using [3.1], while the TD analysis (intra- and inter-tag CC_{max} distributions) is performed using [3.2]. Furthermore, in both the FD and TD analyses, an asymmetrical double sigmoid (ADS) function (Hjelt *et al.* 1999) is used to fit the inter-tag distributions, while a Gaussian (G) probability density function is used to fit the intra-tag distributions.

Figure 6.12 illustrates the similarity analysis for the inkjet-printed PET three-square check-patterned tags, where the FD analysis (Figure 6.12(a)) presents the intra- and inter-tag CS distributions and the TD analysis (Figure 6.12(b)) presents the intra- and inter-tag CC_{max} distributions.

It can be observed from the insets of Figure 6.12(a) and (b) that the intra- and inter-tag distributions present a slight margin between them in both the FD and TD analyses. The error rates are calculated from the actual similarity distributions as well as from fitted probability density functions, as shown in Figure 6.12(c). For the actual similarity distributions, the calculated PE for the FD analysis is $PE_{FD}^{dist} = 0.41\%$ and the calculated PE for the TD analysis is $PE_{TD}^{dist} = 0.33\%$ (see the lines without markers in Figure 6.12(c)). If we use the fitted probability density functions, the calculated PE for FD analysis is $PE_{FD}^{fit} = 0.09\%$ and the calculated PE for TD analysis is $PE_{TD}^{fit} = 0.04\%$ (see the lines with markers in Figure 6.12(c)).

It can be observed that the curves of the error rates from the actual similarity distributions for both FD and TD analyses appear deterministic, as shown in Figure 6.12(c). This behavior is due to the low resolution (i.e. the limited number of samples) of the distributions.

Similarly, for the inkjet-printed PET five-square check-patterned tags, the similarity analysis is presented in Figure 6.13, where the intra- and inter-tag CS distributions for the FD analysis are presented in Figure 6.13(a) and the intra- and inter-tag CC_{max} distributions for the TD analysis are presented in Figure 6.13(b). It can be observed from the insets of Figure 6.13(a) and (b) that the margins between the intra- and inter-tag distributions are larger than the margins between the intra- and inter-tag distributions for three-square check-patterned tags (see the insets of Figure 6.13(a) and (b)). The error rates from the actual similarity distributions cannot be estimated precisely due to the absence of overlaps between the PFP and PFN curves (the line without markers) in both FD and TD analyses, as shown in Figure 6.13(c). The reason for such absence of the overlaps is the low resolution of the distributions (i.e. the limited number of samples). From the broken edges (see the black long-dashed line in Figure 6.13(c)), a rough estimate of the probabilities of error (PEs) for both FD and TD analyses (PE_{FD}^{dist} and PE_{TD}^{dist}) is in the range of 10^{-3}–10^{-2} (i.e. 0.01–0.1%). From the fitted probability density functions, the calculated PE for FD analysis is PE_{FD}^{fit} = 0.013% and the calculated PE for TD analysis is PE_{TD}^{fit} = 0.003% (see the lines with markers in Figure 6.13(c)).

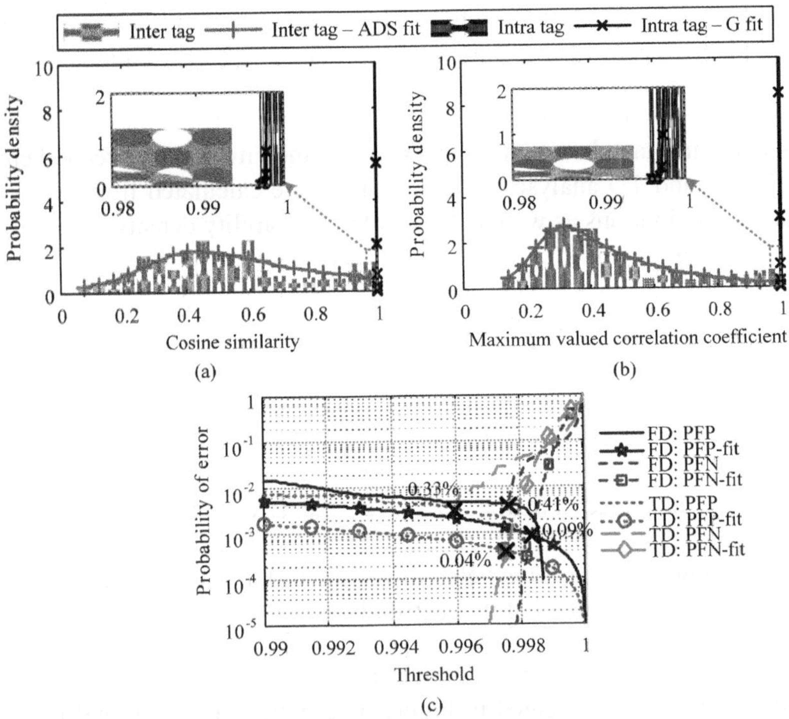

Figure 6.12. *Similarity analysis for the inkjet-printed PET three-square check-patterned tags with d_s = 70.7 μm measured using the VNA-based chipless reader. (a) Intra- and inter-tag cosine similarity distributions for FD analysis. (b) Intra- and inter-tag maximum valued correlation coefficient distributions for TD analysis. (c) Probabilities of false positive and false negative for both FD and TD analyses. For a color version of this figure, see www.iste.co.uk/ali/RFID.zip*

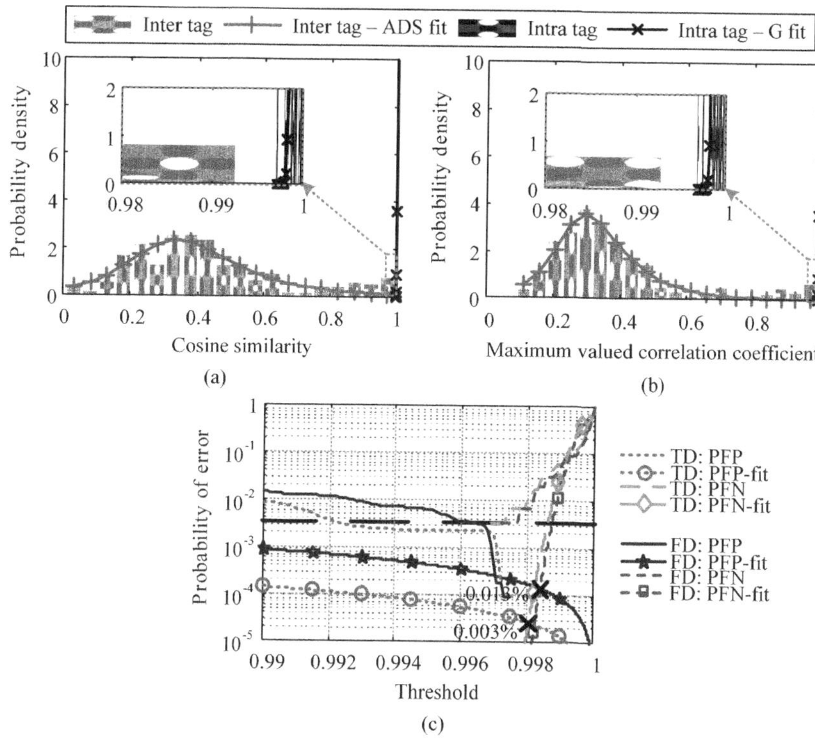

Figure 6.13. *Similarity analysis for the inkjet-printed PET five-square check-patterned tags with d_s = 70.7 μm measured using the VNA-based chipless reader. (a) Intra- and inter-tag cosine similarity distributions for FD analysis. (b) Intra- and inter-tag maximum valued correlation coefficient distributions for TD analysis. (c) Probabilities of false positive and false negative for both FD and TD analyses. For a color version of this figure, see www.iste.co.uk/ali/RFID.zip*

The central tendency, for example, the mean value of the intra-tag distributions of both the FD and TD analyses, is close to unity (see the insets of Figures 6.12 and 6.13), which validates the repeatability of each tag. A slight spread of the intra-tag distributions below the unity is due to the uncertainties (systematic error and random error) from the measurement bench.

Table 6.2 shows an evolution of the PE in comparison to the d_s for the three-square check-patterned C-folded designs. We printed populations of 30 PET tags with three values of d_s: 21.2 μm, 42.4 μm and 70.7 μm.

The three-square check-patterned tags with $d_s = 70.7$ µm have already been characterized in Figure 6.12. For other two populations (i.e. associated with two values of the d_s of 21.2 µm and 42.4 µm), the free space measurements are performed using the measurement setup shown in Figure 6.8 with exactly the same equipment and parameters (e.g. VNA, antennas, cables, distances, output power, frequency sweeping, number of points). The S_{11} of each tag is measured five times with removing and repositioning manner. The rest of the post-treatment on the measured signals (background normalization and time windowing along with the chosen parameters for time window) is the same as discussed above in this section. Similarly, the distributions corresponding to the FD and the TD are calculated using [3.1] and [3.2], respectively. Finally, the PEs are calculated using the similarity distributions PE_{FD}^{dist} and PE_{TD}^{dist} (not the fitted probability density functions). From the results in Table 6.2, a trend of increasing performance as d_s increases can be observed.

d_s (µm)	Probability of error (%)	
	FD	**TD**
21.2	6	5.46
42.4	3	2.24
70.7	0.41	0.33

Table 6.2. *Evolution of the probability of error in comparison to the vertex-to-vertex gap among the squares for the three-square check-patterned C-folded design*

6.4. Authentication using IR-UWB chipless reader

For a cheap practical solution, an in-house built IR-UWB chipless RFID reader (Garbati *et al.* 2018) is also used. This IR-UWB chipless reader is developed using off-the-shelf available components that result in a cheap device. The cost of this reader is comparable with the classical RFID reader used in logistics and supply chain management. The band of operation of this reader is 3.1–10.6 GHz. For the free space measurements using the IR-UWB chipless reader, a co-polarization bistatic configuration inside the anechoic environment is used, as shown in Figure 6.14, where the antennas, the spacing between the antennas and the distance between the tag and antennas are exactly the same as discussed in the measurement setup with VNA, as shown in Figure 6.8.

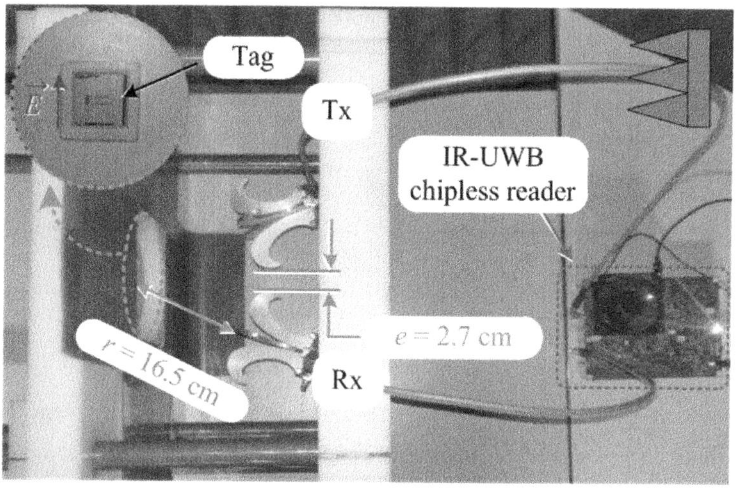

Figure 6.14. *Measurement setup in an anechoic environment with the IR-UWB chipless reader. For a color version of this figure, see www.iste.co.uk/ali/RFID.zip*

The five-square check-patterned tags outperform the three-square check-patterned tags in the VNA-based chipless reader case (see section 6.3). For this reason, the five-square check-patterned tags have been chosen to analyze the performance with a low-cost IR-UWB chipless reader in this section. Like the previous measurement manner, each tag is measured five times successively by removing and repositioning the tag at its position during each measurement trial to check the repeatability. The measured parameter is the transmission coefficient S_{21} due to the reader configuration. The post-processing (i.e. background normalization and time windowing) is also applied. For instance, the time windowing for the S_{21} of the first tag is shown in Figure 6.15. The start time of the time window (13 ns) along the window length (7 ns) is shown in Figure 6.15. Here, the parameters of the post-processing are optimized for this specific signal (i.e. limited to 7 ns due to the wire reflections), as the IR-UWB reader is totally different from the VNA, where IR-UWB uses a UWB pulse in the TD. The non-windowed TD signal (i.e. measured using the measurement configuration shown in Figure 6.14) exhibits 3789 data points. After time windowing, the windowed TD signal exhibits 700 data points by neglecting the padded zeros. The padded zeros do not contribute to the similarity while comparing the signals. The windowed FD signal exhibits 380 points.

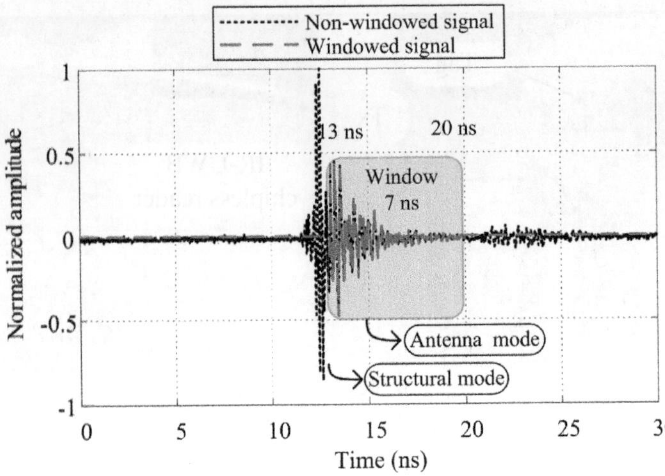

Figure 6.15. *Time windowing to discard the structural mode and to extract the antenna mode. The start time of window is 13 ns and the duration of the time window is 7 ns. For a color version of this figure, see www.iste.co.uk/ali/RFID.zip*

The first measurement of S_{21} in the form of windowed FD filtered responses for all the 30 tags based on five-square check-patterned C-folded scatterer (Figure 6.7(b)) is shown in Figure 6.16. Here, too, the randomness that happened during the inkjet printing process can be observed from the randomness of the position of the peak apexes corresponding to the f_rs of the tags.

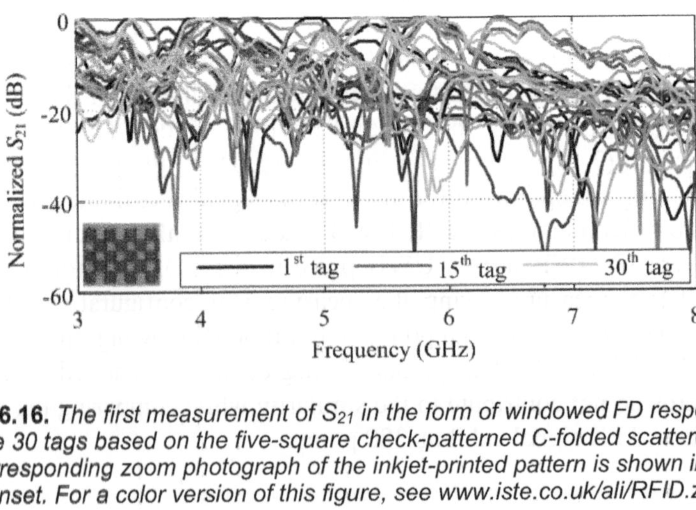

Figure 6.16. *The first measurement of S_{21} in the form of windowed FD responses for all the 30 tags based on the five-square check-patterned C-folded scatterer. The corresponding zoom photograph of the inkjet-printed pattern is shown in the inset. For a color version of this figure, see www.iste.co.uk/ali/RFID.zip*

On the other hand, Figure 6.17 shows an example of the repeatability of the measurement system by presenting the repetitive measurements of S_{21} in the form of windowed filtered FD responses for the second sample from the five-square check-patterned C-folded scatterer tag using the IR-UWB chipless reader. Here, it can be observed that the repeatability is lower than the case of the VNA-based reader (Figure 6.11). This is because of the low accuracy of the IR-UWB chipless RFID reader.

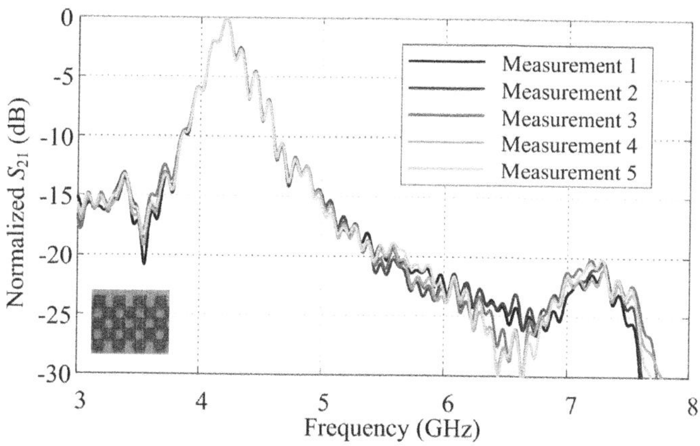

Figure 6.17. *Five repetitive measurements of S_{21} in the form of windowed FD responses for the second sample from the five-square check-patterned C-folded scatterer tags using the IR-UWB chipless reader. The corresponding zoom photograph of the inkjet-printed pattern is shown in the inset. For a color version of this figure, see www.iste.co.uk/ali/RFID.zip*

Figure 6.18 presents the similarity analyses for inkjet-printed PET five-square check-patterned tags measured using the IR-UWB chipless reader. Here, too, the FD analysis (Figure 6.18(a)) presenting the intra- and inter-tag CS distributions is calculated using [3.1], and the TD analysis (Figure 6.18(b)) presenting the intra- and inter-tag CC_{max} distributions is calculated using [3.2]. The error rates are calculated from the actual similarity distributions as well as from fitted probability density functions, as shown in Figure 6.18(c). For the actual similarity distributions, the calculated PE for the FD analysis is $PE_{FD}^{dist} = 2.3\%$ and the calculated PE for the TD analysis is $PE_{TD}^{dist} = 2.7\%$ (see the lines without markers in Figure 6.18(c)). If we use the fitted probability density functions, the calculated PE for the FD analysis is $PE_{FD}^{fit} = 1.1\%$ and the calculated PE for the TD analysis is $PE_{TD}^{fit} = 3\%$ (see the lines with markers in

Figure 6.18(c)). It can be observed that the $PE_{TD}^{fit} > PE_{TD}^{dist}$, which is due to the overestimation of distribution from the fitting function (i.e. ADS).

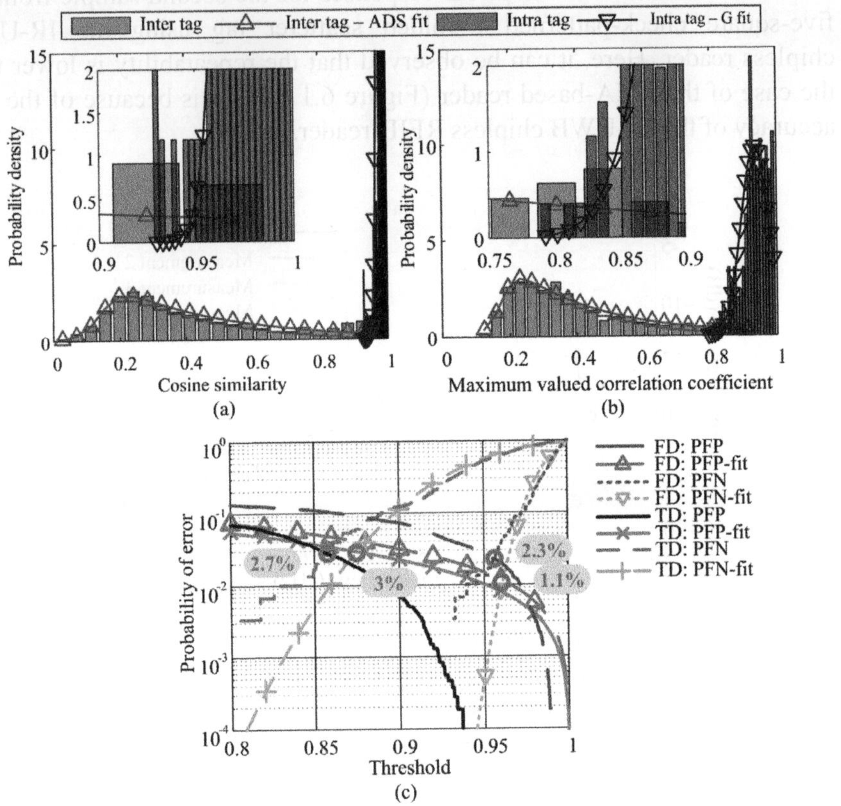

Figure 6.18. *Similarity analysis for the inkjet-printed PET five-square check-patterned tags with d_s = 70.7 μm measured using the IR-UWB chipless reader. (a) Intra- and inter-tag cosine similarity distributions for FD analysis. (b) Intra- and inter-tag maximum valued correlation coefficient distributions for TD analysis. (c) Probabilities of false positive and false negative for both FD and TD analyses. For a color version of this figure, see www.iste.co.uk/ali/RFID.zip*

It can be observed that for the IR-UWB chipless reader (Figure 6.18), the PE_{TD}^{dist} or the PE_{TD}^{fit} are larger than the PE_{FD}^{dist} or the PE_{FD}^{fit}. While for the VNA-based chipless reader (Figure 6.12 or Figure 6.13), the PE_{TD}^{dist} or the PE_{TD}^{fit} are less than the PE_{FD}^{dist} or the PE_{FD}^{fit}. Furthermore, it can also be observed that for the IR-UWB chipless reader (Figure 6.18), the spread of the inter-tag distribution in the FD is larger than the spread of the inter-tag distribution in the TD, while the spread of the intra-tag distribution in FD is smaller than the spread of the inter-tag distribution in TD. The reason for these occurrences is that in the IR-UWB chipless reading system, the signal acquisition is in the TD, while in the VNA-based chipless reading configuration, the signal acquisition is in the FD. The TD signal from the cheap IR-UWB reader is less accurate than the FD signal from highly costly and highly accurate VNA. This may give rise to the low PE_{FD} than to the PE_{TD} and variation in the spread among the intra- and inter-tag distributions.

For five-square check-patterned chipless tags, the PE_{FD}^{dist} and PE_{TD}^{dist} are in the range of 10^{-3}–10^{-2} (i.e. 0.01–0.1%) with the VNA. Such levels of PE are quite insignificant and comparable to the average PEs for the fingerprint verification system ranging from 2.07% to 4.03%, as found in Cappelli *et al.* (2006), and the best average PE of 0.05%, among different fingerprint evaluation campaigns, as found in Maltoni *et al.* (2009, Chap. 4). Furthermore, the average PEs for the fingerprint verification system ranging from 2.07% to 4.03%, as found in Cappelli *et al.* (2006), are also comparable to the results obtained using the IR-UWB chipless reader (e.g. $PE_{FD}^{dist} = 2.3\%$ and $PE_{TD}^{dist} = 2.7\%$).

The obtained results show that the proposed technique presents the acceptable performance with the highly accurate VNA-based reader as compared to the various fingerprint evaluation campaigns found in the literature. On the other hand, the obtained performance with the IR-UWB chipless reader is comparable to the fingerprint evaluation campaigns found in the literature. The performance of the IR-UWB chipless reader is 10 times lower than that of the VNA-based reader, but the cost of the IR-UWB chipless reader is a hundred times less than that of the VNA-based reader. The proposed authentication method has a significant advantage of the possibility of hiding the chipless RFID tag entirely by virtue of the RF.

We can say that the fitting function (i.e. ADS) applied to the inter-tag distributions underestimates or overestimates them near tails (between 0.8 and 1). We provide a simple method to model the system, as we have seen in Figure 6.13(c), the curves of the error rate from the actual similarity distributions are not overlapping. On the other hand, we have also provided the results using the actual distributions. However, the more complex fitting method such as using multi-probability density functions can be used for the precise modeling of the system.

We can also ask on what criteria a square geometry ■ is chosen to make patterns in this chapter. For this purpose, symmetrical shapes are good candidates, because symmetrical shapes provide a vertex-to-vertex (or point-to-point) connectivity in the pattern. Such vertex-to-vertex connected patterns can exploit the randomness inherent in the inkjet printing significantly as compared to the non-vertex-to-vertex connected patterns. In the context of other possible candidates for the optimized chipless tag, a diamond geometry ♦ may be equivalent to the square geometry ■, as both geometries are polygons with four vertices and symmetrical shapes. On the other hand, for a circle geometry ●, we have testified that it provides sometimes well-connected patterns or well-disconnected patterns, due to the arc-to-arc nature of the pattern. This behavior leads the circle pattern as a weak candidate for an optimized design of chipless tags for authentication. For more randomness vulnerable designs, an hourglass geometry (based on a triangular geometry) ⧗ or an octagram geometry ✸ can be used. The former has five vulnerable points and the latter has eight vulnerable points. These geometries may be complex in design. One other possibility is to use a gradient pattern in the input digital mask to realize the optimized chipless tags. The results related to this design are not discussed in the chapter because of the lower performance as compared to square check patterns. In summary, we have used a simple square geometry to explain our idea in an uncomplicated way.

6.5. Conclusion

The opportunity to exploit the process randomness inherent in inkjet printing for the generation of unique EM signals was discussed in this chapter. These unique signals were then used to provide a highly secure authentication solution. The design of a conventional C-folded scatterer was optimized by introducing discrete square patterns along the arms of the C-folded scatterer to elevate its sensitivity to the inkjet printing. Then, the proposal was validated by showing the performance of the optimized sensitive check-patterned C-folded chipless tags realized on a PET printable film using a highly accurate VNA-based chipless reader as well as using a low-cost IR-UWB chipless reader. An evolution of the PE had also been presented in comparison to the vertex-to-vertex gap d_s among the squares of the check pattern. The achieved PE is comparable to the various fingerprint evaluation campaigns found in the literature. The advantage of the proposed chipless authentication is that in contrast to the classical image-based authentication approach (like a footprint), the chipless tag can be totally hidden (potentially without performance degradation) for confidentiality and for the product integrity of high-end luxury products.

Conclusion

This book was focused on taking the next step with the aim of developing chipless tags for authentication applications. The concept of chipless RFID was extended to authentication where each tag had to present a unique signature that could never be reproduced even if someone tried to copy the tag. For this purpose, natural randomness (i.e. inherent in the fabrication process) along the dimensional parameters of resonators was used. Such natural randomness can produce unique EM signatures that can be used for authentication.

A methodology to characterize chipless RFID tags for the authentication application was presented. For this purpose, procedures to conduct authentication and statistical analyses were presented. A proof of concept was demonstrated by using two technologies: PCB and inkjet printing. Chipless tag discrimination based on the level of similarity was presented in both the frequency and time domains.

Then, for the extraction of aspect-independent parameters, a fast and efficient spectrogram approach was presented. These aspect-independent parameters are the main information that possesses the uniqueness of the tags originating from the natural process variations, because the parameters are independent of the aspects of the measurement setup. The extraction of complex natural frequency(ies) using the spectrogram was never performed earlier in the field of frequency-coded chipless RFID. With an operation of a single measurement, the proposed technique is very promising for the practical implementation of the chipless RFID technology, as it is computationally less expensive due to the inherent fast property of fast

Fourier transform. Thus, the proposed technique requires fewer resources and efforts.

The chipless authentication using PCB was demonstrated. First, it was shown that the four-coupled C-folded scatterer-based chipless tag was a better choice than the single C-folded scatterer-based chipless tag due to its high Q. To show the minimum detectable dimensional variation, three groups of tags (quad C-folded scatterers tags) were fabricated, which showed distinct arms' lengths that mimicked the process variations occurring in the fabrication process. A second-order bandpass filter model was used to describe the antenna mode of a C-folded resonator. Owing to this proposed model along with the aspect-independent parameters, it was demonstrated that two groups of the tags showing a measured mean variation of approximately 25 μm along the arms' length could be discriminated with a success rate greater than 99.9% even in a realistic outdoor environment. A good agreement was shown between the theoretical model and the experimental results. According to the validated model, the proposed discrimination technique could potentially differentiate two close structures whose dimensions varied in the accuracy range of the classical manufacturing technique used to realize cheap tags. Finally, natural dimensional variations in the design of C-folded tags were analyzed for the authentication applications. For this purpose, quad-scatterer tags were chosen because of their sharp slope dissimilarity. Chipless tags were realized two times intermittently, where each realization consisted of 45 tags. The two different realizations shared the same company, the same PCB technology, but a different film mask, in order to ensure the natural dimensional randomness. The similarity analyses were conducted within each realization and between two intermittent realizations. The probability of error achieved was comparable to the various fingerprint evaluation campaigns found in the literature. The existence of the natural dimensional variations was also confirmed by the microscopic dimensional analysis of two intermittent realizations. Then, the minimum and maximum changes between the intermittent realizations were also linked to the similarity change to show an intuitive range. It was found that the existing natural variations were sufficient to produce similarity change to be used for authentication applications. Finally, the technique was generalized to reduce the probability of error to a significant level.

Finally, the opportunity to exploit the process randomness inherent in inkjet printing for the generation of unique electromagnetic signals was discussed. These unique signals were then used to provide a highly secure authentication solution. The design of a conventional C-folded scatterer was optimized by introducing discrete square patterns along the arms of the C-folded scatterer, in order to elevate its sensitivity to inkjet printing. Then, the proposal was validated by showing the performance of optimized sensitive check-patterned C-folded chipless tags fabricated on a PET printable film using a highly accurate VNA-based chipless reader and a low-cost IR-UWB chipless reader. An evolution of PE was also presented in comparison to the vertex-to-vertex gap among the squares of the check pattern. The PE achieved was comparable to the various fingerprint evaluation campaigns found in the literature.

As a first step towards the realistic environment operation of the proposed chipless authentication method, dimensional variations ΔL can be purposely applied to the chipless tags. The magnitude of such purposely applied dimensional variations ΔL can be much larger than the fabrication tolerance in order to make them easily detectable. This act of purposely applying dimensional variations ΔL will result in a less secure authentication technique. Therefore, the proposed authentication technique shows a trade-off between security level and realistic operation.

Appendices

Appendix A

Calculation of the Effective Permittivity of a Coplanar Stripline

The effective permittivity ε_{eff} for a coplanar stripline C-folded scatterer with finite substrate thickness (as shown in Figure A.1) can be calculated from the following expression (Garg *et al.* 2013, Chap. 7):

$$\varepsilon_{eff} = 1 + \frac{\varepsilon_r - 1}{2} \frac{K'(k_{10})}{K(k_{10})} \frac{K(k_1)}{K'(k_1)}, \quad [A.1]$$

where ε_r corresponds to the dielectric constant of the substrate.

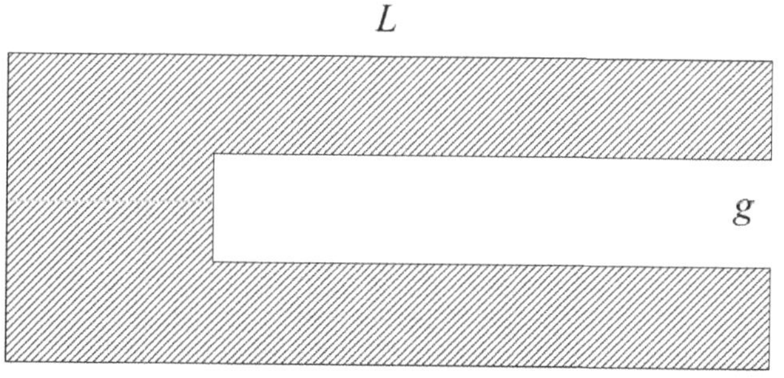

Figure A.1. *Coplanar stripline C-folded scatterer with finite substrate thickness*

The parameters k_{10} and k_1 are calculated as

$$k_{10} = \frac{\tanh\left(\frac{\pi g}{4h}\right)}{\tanh\left(\frac{\pi}{4}\frac{2w+g}{h}\right)}, \quad [A.2]$$

$$k_1 = \frac{g}{g+2w}, \quad \text{[A.3]}$$

where h is the substrate thickness. The parameters w and g correspond to the width and spacing between the coplanar striplines (see Figure A.1). The ratio $K(k)/K'(k)$ for a variable k is calculated as follows:

$$\frac{K(k)}{K'(k)} = \frac{pi}{ln[2(1+\sqrt{k'})/(1+\sqrt{k'})]} \quad \text{for } 0 \leq k \leq 0.707 \quad \text{[A.4a]}$$

$$\frac{K(k)}{K'(k)} = \frac{1}{\pi} ln[2(1+\sqrt{k})/(1+\sqrt{k})] \quad \text{for } 0.707 \leq k \leq 1 \quad \text{[A.4b]}$$

$$k' = \sqrt{1-k^2} \quad \text{[A.4c]}$$

Appendix B

Measurement Setup Inside an Anechoic Chamber

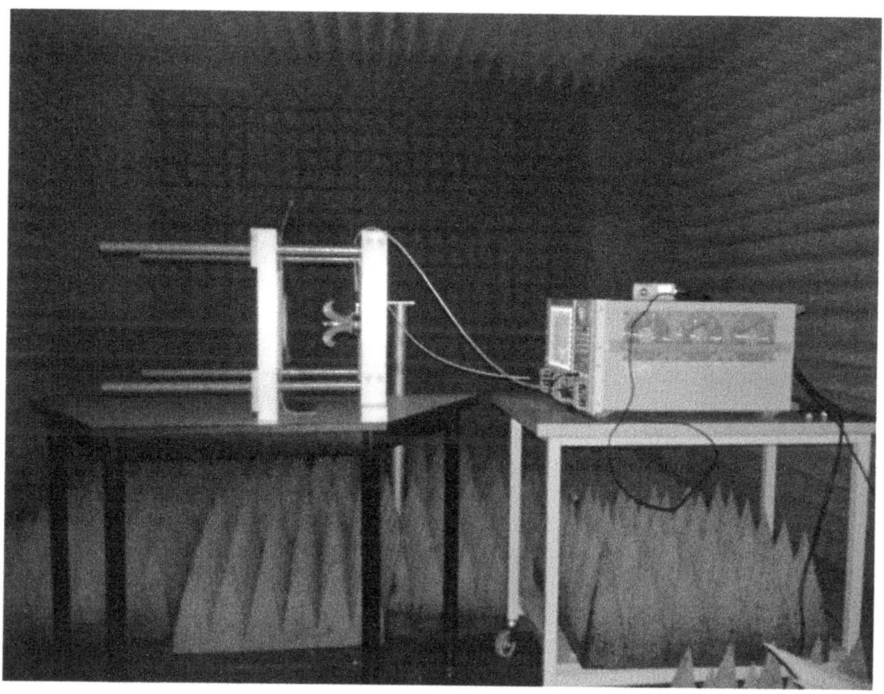

Figure B.1. *Measurement setup inside an anechoic chamber. For a color version of this figure, see www.iste.co.uk/ali/RFID.zip*

Appendix C

Matrix Pencil Method

The TD backscattered field from a chipless tag can be written as

$$y(t) \approx \sum_{i=1}^{M} R_i z_i, \qquad [C.1]$$

where M is the number of scatterers in the chipless RFID tag; R_i is the residue; $z_i = e^{s_i t}$, $s = -\sigma + j\omega$ (with σ being the damping factor and ω the angular frequency) is the CNR of a second-order pole; and t is the time.

For the implementation of the total least-squares MPM, a data matrix [Y] is constructed using the TD signal $y(t)$:

$$[\mathrm{Y}] = \begin{bmatrix} y(0) & y(1) & \cdots & y(L) \\ y(1) & y(2) & \cdots & y(L+1) \\ \vdots & \vdots & & \vdots \\ y(N-L-1) & y(N-L) & \cdots & y(N-1) \end{bmatrix}_{(N-L)\times(L+1)} \qquad [C.2]$$

where N is the total data samples in $y(t)$ and L is the pencil parameter that is useful to eliminate the effect of noise from data. The optimum value of L should be chosen between $N/3$ and $N/2$.

The MPM uses singular value decomposition to reduce the dimensionality of the data [Y]:

$$[\mathrm{Y}] = [\mathrm{U}][\Sigma][\mathrm{V}]^{\mathrm{H}}, \qquad [C.3]$$

where [U] and [V] are unitary matrices; [Σ] is a diagonal matrix comprising the singular values of [Y]; and the operator ^{H} corresponds to the Hermitian transpose.

In the noisy case, [Y] does not contain exactly M non-zero eigenvalues corresponding to the M poles of the system. The singular values of the data [Y] are then used to define the desired number of poles *M* by comparing the ratio of each singular value δ to the largest singular value δ_{max} with a threshold value:

$$\frac{\delta}{\delta_{max}} = 10^{-p}, \qquad [C.4]$$

where *p* is the number of significant decimal digits.

Then, the filtering of [V] can be performed by extracting only M vectors:

$$[V'] = [v_1, v_2, \dots, v_M]. \qquad [C.5]$$

Using [V′], two new submatrices can be defined as follows:

$$[V'] = \begin{bmatrix} l_1 \\ V_1' \end{bmatrix}_{(L+1)\times M} \qquad [C.6]$$

$$[V'] = \begin{bmatrix} V_2' \\ l_{L+1} \end{bmatrix}_{(L+1)\times M} \qquad [C.7]$$

where l_1 is the *i*th line of [V′].

Finally, using the total least-squares approach (Sarrazin *et al.* 2014), the complex poles can be obtained from the eigenvalues of $\{[V_1']^H\}^+ [V_2']^H$, where $\{[V_1']^H\}^+$ is the Moore–Penrose pseudo-inverse of $[V_1']^H$. Using the extracted complex poles, TD signals can be constructed and the residues can be calculated from [C.1].

For more details on the algorithm of the MPM for calculating the complex poles and residues, see Sarkar and Pereira (1995).

References

Ali, Z., Barbot, N., Siragusa, R., Hely, D., Bernier, M., Garet, F., and Perret, E. (2018a). Chipless RFID tag discrimination and the performance of resemblance metrics to be used for it. In *Proceedings of the IEEE MTT-S International Microwave Symposium (IMS)*, Philadelphia, PA, USA, 363–366.

Ali, Z., Barbot, N., Siragusa, R., Perret, E., Hely, D., Bernier, M., and Garet, F. (2018b). Detection of minimum geometrical variation by free-space-based chipless approach and its application to authentication. *IEEE Microwave and Wireless Components Letters*, 28, 323–325.

Ali, Z., Perret, E., Barbot, N., Siragusa, R., Hely, D., Bernier, M., and Garet, F. (2019). Detection of natural randomness by chipless RFID approach and its application to authentication. *IEEE Transactions on Microwave Theory and Techniques*, 67, 3867–3881.

Ali, Z., Perret, E., Barbot, N., Siragusa, R., Hely, D., Bernier, M., and Garet, F. (2020). Authentication using metallic inkjet-printed chipless RFID tags. *IEEE Transactions on Antennas and Propagation*, 68(5), 4137–4142, https://doi.org/10.1109/TAP.2019.2948740.

Allen, J. (1982). Applications of the short time Fourier transform to speech processing and spectral analysis. In *Proceedings of the 7th IEEE International Conference on Acoustics, Speech, and Signal Processing*, 1012–1015.

Alves, A.A.C., Spadoti, D.H., and Bravo-Roger, L.L. (2018). Optically controlled multiresonator for passive chipless tag. *IEEE Microwave and Wireless Components Letters*, 28, 467–469.

Amin, E.M., Bhuiyan, M.S., Karmakar, N.C., and Winther-Jensen, B. (2014). Development of a low cost printable chipless RFID humidity sensor. *IEEE Sensors Journal*, 14, 140–149.

Anee, R. and Karmakar, N.C. (2013). Chipless RFID tag localization. *IEEE Transactions on Microwave Theory and Techniques*, 61, 4008–4017.

Auslander, J.D. and Berson, W. (1997). Bar code printing and scanning using wax based invisible fluorescent inks. US5693693A.

Barahona, M., Betancourt, D., and Ellinger, F. (2014). Decoding of multiple same-coded in-line placed chipless RFID tags. In *Proceedings of the IEEE Conference on Antenna Measurements & Applications (CAMA)*, 1–4, https://doi.org/10.1109/CAMA.2014.7003353.

Barahona, M., Betancourt, D., and Ellinger, F. (2016a). Comparison of UWB chipless tags on flexible substrates fabricated using either aluminum, copper or silver. In *Proceedings of the IEEE-APS Topical Conference on Antennas and Propagation in Wireless Communications (APWC)*, 78–81, https://doi.org/10.1109/APWC.2016.7738123.

Barahona, M., Betancourt, D., and Ellinger, F. (2016b). Using UWB IR radar technology to decode multiple chipless RFID tags. In *Proceedings of the Conference on Ubiquitous Wireless Broadband (ICUWB)*, 1–6.

Barahona, M., Betancourt, D., and Ellinger, F. (2016c). On the decoding of equiprobable UWB chipless RFID tags using a minimum distance detector. In *Proceedings of ISAP*, Okinawa, Japan, p. 2.

Barahona, M., Betancourt, D., Ellinger, F., Haase, K., Schmidt, G.C., and Hübler, A.C. (2017). Automatic IR UWB chipless RFID system for short range applications. In *Proceedings of the IEEE-APS Topical Conference on Antennas and Propagation in Wireless Communications (APWC)*, 274–279.

Barbot, N. and Perret, E. (2017). Gesture recognition with the chipless RIFD technology. In *Proceedings of the XXXIInd General Assembly Scientific Symposium International Union Radio Science (URSI GASS)*, 1–3.

Barbot, N. and Perret, E. (2018). A chipless RFID method of 2D localization based on phase acquisition. *Journal of Sensors*, 2018, 1–6, https://doi.org/10.1155/2018/7484265.

Batchelor, J.C., Parker, E.A., Miller, J.A., Sanchez-Romaguera, V., and Yeates, S.G. (2009). Inkjet printing of frequency selective surfaces. *Electronic Letters*, 45, 7–8.

Baum, C.E., Rothwell, E.J., Chen, K.M., and Nyquist, D.P. (1991). The singularity expansion method and its application to target identification. In *Proceedings of the IEEE*, 79, 1481–1492.

Berning, P.H. and Phillips, R.W. (1990). Thin film optical variable article and method having gold to green color shift for currency authentication. US4930866A.

Bertoncini, C., Rudd, K., Nousain, B., and Hinders, M. (2012). Wavelet fingerprinting of radiofrequency identification (RFID) tags. *IEEE Transactions on Industrial Electronics*, 59, 4843–4850.

Betancourt, D., Nair, R., Haase, K., Schmidt, G., Bellmann, M., Höft, D., Hübler, A., and Ellinger, F. (2015). Square-shape fully printed chipless RFID tag and its applications in evacuation procedures. In *Proceedings of the 9th European Conference on Antennas and Propagation (EuCAP)*, 1–5.

Betancourt, D., Haase, K., Hübler, A., and Ellinger, F. (2016). Bending and folding effect study of flexible fully printed and late-stage codified octagonal chipless RFID tags. *IEEE Transactions on Antennas and Propagation*, 64, 2815–2823.

Betancourt, D., Barahona, M., Haase, K., Schmidt, G., Hubler, A., and Ellinger, F. (2017). Design of printed Chipless-RFID tags with QR-code appearance based on genetic algorithm. *IEEE Transactions on Antennas and Propagation*, 65, 2190–2195.

Blischak, A.T. and Manteghi, M. (2011). Embedded singularity chipless RFID tags. *IEEE Transactions on Antennas and Propagation*, 59, 3961–3968.

Borgese, M., Dicandia, F.A., Costa, F., Genovesi, S., and Manara, G. (2017). An inkjet printed chipless RFID sensor for wireless humidity monitoring. *IEEE Sensors Journal*, 17, 4699–4707.

Borghetti, M., Ghittorelli, M., Sardini, E., Serpelloni, M., and Torricelli, F. (2016). Electrical characterization of PEDOT:PSS strips deposited by inkjet printing on plastic foil for sensor manufacturing. *IEEE Transactions on Instrumentation and Measurement*, 65, 2137–2144.

Boutant, Y., Becker, J.M., and Fournel, T. (2007). Method for extracting random signatures from a material element and method for generating a decomposition base to implement the extraction method. CA2953689 (A1).

Bradley, R.A. and Witzman, M.R. (2001). Color shifting thin film pigments. US6243204B1.

Brock, D.L. (2001). *The Electronic Product Code (EPC) A Naming Scheme for Physical Objects*. Massachusetts Institute of Technology, Cambridge, MA, USA.

Brogger, B., Esterberg, J., and Benderly, D. (2004). Non-toxic biodegradable microtaggants. US20040029295A1.

Buchanan, J.D.R., Cowburn, R.P., Jausovec, A.V., Petit, D., Seem, P., Xiong, G., Atkinson, D., Fenton, K., Allwood, D.A., and Bryan, M.T. (2005). Forgery: 'Fingerprinting' documents and packaging. *Nature*, 436, 475.

Cappelli, R., Maio, D., Maltoni, D., Wayman, J.L., and Jain, A.K. (2006). Performance evaluation of fingerprint verification systems. *IEEE Transactions on Pattern Analysis and Machine Intelligence*, 28, 3–18.

Carnahan, R.D. and Hou, S.L. (1977). Ink jet technology. *IEEE Transactions on Industry Applications*, IA-13, 95–104.

Carro-Temboury, M.R., Arppe, R., Vosch, T., and Sørensen, T.J. (2018). An optical authentication system based on imaging of excitation-selected lanthanide luminescence. *Science Advances*, 4, e1701384.

Cass, T.A. and Marimont, D.H. (1997). Binary glyph codes based on color relationships. US5684885A.

Chaudhry, P. and Zimmerman, A. (2013). *Protecting Your Intellectual Property Rights: Understanding the Role of Management, Governments, Consumers and Pirates*. Springer, New York, NY, 7–31, https://doi.org/10.1007/978-1-4614-5568-4_2.

Choi, S.H., Yang, B., Cheung, H.H., and Yang, Y.X. (2015). RFID tag data processing in manufacturing for track-and-trace anti-counterfeiting. *Computers in Industry*, 68, 148–161.

Christofferson, D.F., Anderson, E.J., Geres, J.C., and Graushar, W.T. (2009). Resonator use in the print field. US7506813B2.

Costa, F., Borgese, M., Gentile, A., Buoncristiani, L., Genovesi, S., Dicandia, F.A., Bianchi, D., Monorchio, A., and Manara, G. (2018a). Robust reading approach for moving chipless RFID tags by using ISAR processing. *IEEE Transactions on Microwave Theory and Techniques*, 66, 2442–2451.

Costa, F., Gentile, A.F., Genovesi, S., Buoncristiani, L., Lázaro, A., Villarino, R.M., and Girbau, D. (2018b). A depolarizing chipless RF label for dielectric permittivity sensing. *IEEE Microwave and Wireless Components Letters*, 28, 371–373.

Cruikshank, D.N.C., Merry, T., and Suzzarini, L.M.-F. (2006). Security document with ultraviolet authentication security feature. EP1719637A2.

Danev, B., Heydt-Benjamin, T.S., and Capkun, S. (2009). Physical-layer identification of RFID devices. In *Proceedings of the 8th Conference on USENIX Security Symposium*, 199–214.

Danev, B., Capkun, S., Masti, R.J., and Benjamin, T.S. (2012). Towards practical identification of HF RFID devices. *ACM Transactions on Information and System Security (TISSEC)*, 15, 7.

Deepu, V., Vena, A., Perret, E., and Tedjini, S. (2010). New RF identification technology for secure applications. In *Proceedings of the IEEE International Conference on RFID-Technology and Applications*, Guangzhou, China, 159–163.

DeJean, G. and Kirovski, D. (2007). Radio frequency certificates of authenticity. US20070005367A1.

DeJean, G., Lakafosis, V., Traille, A., Lee, H., Gebara, E., Tentzeris, M., and Kirovski, D. (2011). RFDNA: A wireless authentication system on flexible substrates. In *Proceedings of the IEEE 61st Electronic Components and Technology Conference (ECTC)*, Lake Buena Vista, FL, USA, 1332–1337.

Deng, F., He, Y., Li, B., Song, Y., and Wu, X. (2018). Design of a slotted chipless RFID humidity sensor tag. *Sensors and Actuators B: Chemical*, 264, 255–262.

Donelli, M. (2018). An RFID-based sensor for masonry crack monitoring. *Sensors*, 18, 4485, https://doi.org/10.3390/s18124485.

Duineveld, P.C. (2003). The stability of ink-jet printed lines of liquid with zero receding contact angle on a homogeneous substrate. *Journal of Fluid Mechanics*, 477, 175–200, https://doi.org/10.1017/S0022112002003117.

Eastwood, I., Cartwright, J., Dorland, E., and Saw, M. (2011). Authentication of ingestible products using saccharides as markers. US7972859B2.

Feng, Y., Xie, L., Chen, Q., and Zheng, L. (2015). Low-cost printed chipless RFID humidity sensor tag for intelligent packaging. *IEEE Sensors Journal*, 15, 3201–3208.

Forouzandeh, M. and Karmakar, N.C. (2015). Chipless RFID tags and sensors: A review on time-domain techniques. *Wireless Power Transfer*, 2, 62–77.

Foster, J.J. and Mulcahy, L.T. (2000). Process for making and detecting anti-counterfeit paper. US6045656A.

Fragala, J.S., Loiret-Bernal, C., Zhang, H., Shile, R.R., Rosner, B., and Cruchon-Dupeyrat, S. (2011). Stamps with micrometer- and nanometer-scale features and methods of fabrication thereof. US8069782B2.

Frideling, J.-P. (2006). Method for isotopically tagging products and applications thereof especially in the prevention of counterfeiting. EP1677105A1.

Garbati, M., Siragusa, R., Perret, E., and Halopé, C. (2015). Low cost low sampling noise UWB Chipless RFID reader. In *Proceedings of the IEEE MTT-S International Microwave Symposium (IMS)*, 1–4.

Garbati, M., Ramos, A., Siragusa, R., Perret, E., and Halopé, C. (2016). Chipless RFID reading system independent of polarization. In *Proceedings of the IEEE MTT-S International Microwave Symposium (IMS)*, 1–3.

Garbati, M., Perret, E., Siragusa, R., and Halopè, C. (2017). Toward chipless RFID reading systems independent of tag orientation. *IEEE Microwave and Wireless Components Letters*, 27, 1158–1160.

Garbati, M., Perret, E., and Siragusa, R. (2018). *Chipless RFID Reader Design for Ultra-Wideband Technology: Design, Realization and Characterization.* ISTE Press, London and Elsevier, Oxford.

Gardner, N.A. and Voticky, M.P. (1985). Method of rendering documents resistant to photocopying, and anti-copying paper and ink therefor. US4522429A.

Garg, R., Bahl, I., and Bozzi, M. (2013). *Microstrip Lines and Slotlines*, 3rd edition. Microwave & RF, Artech House.

Gaynor, S.G., McLeod, A.E., Rodig, M.J., Steach, J.K., Collazo, H., Wilson, S.A., Salyer, L.J., Seiler, B.D., and Horton, J. (2017). Fibers with chemical markers used for coding. US9851341B2.

Gaynor, S.G., McLeod, A.E., Wilson, S.A., Collazo, H., Renfro, L.W., Salyer, L.J., Steach, J.K., Rodig, M.J., Seiler, B.D., Horton, J., and Tatum, C. (2018). Fibers with chemical markers and physical features used for coding. US9863920B2.

Glossary Terms IACC [Online]. Available: https://www.iacc.org/resources/glossary-terms (Accessed 21 November 2018).

Golan, R. (2008). Method and apparatus for providing embossed hidden images. US20080134912A1.

Gooch, J., Daniel, B., Abbate, V., and Frascione, N. (2016). Taggant materials in forensic science: A review. *TrAC Trends in Analytical Chemistry*, 83, 49–54. https://doi.org/10.1016/j.trac.2016.08.003.

Gov, S. and Shtrikman, S. (1995). On isotropic scattering of electromagnetic radiation. In *Proceedings of the 18th Convent. Elect. Electron. Eng*, Tel Aviv, Israel, 2.4.4/1–2.4.4/5.

Greene, M. and Hurley, G.W. (2002). Method for forming a radio frequency responsive target and apparatus for verifying the authenticity of same. US6471878 B1.

Greer, P. and Wiklof, C. (2002). Method and apparatus for accessing product information using bar code data. US20020165931A1.

Gröchenig, K. (2001). *Foundations of Time-Frequency Analysis, Applied and Numerical Harmonic Analysis*. Birkhäuser, Boston.

Guillet, A., Vena, A., Perret, E., and Tedjini, S. (2012). Design of a chipless RFID sensor for water level detection. In *Proceedings of the International Symposium on Antenna Technology and Applied Electromagnetics*, 1–4.

Herrojo, C., Mata-Contreras, J., Nunez, A., Paredes, F., Ramon, E., and Martin, F., (2017). Near-field chipless-RFID system with high data capacity for security and authentication applications. *IEEE Transactions on Microwave Theory and Techniques*, 65, 5298–5308.

Hoath, S.D., Jung, S., Hsiao, W.K., and Hutchings, I.M. (2012). How PEDOT:PSS solutions produce satellite-free inkjets. *Organic Electronics*, 13, 3259–3262.

Howland, P. and Foulkes, J.P. (1999). Security paper. US5868902A.

Hu, S., Zhou, Y., Law, C.L., and Dou, W. (2010). Study of a uniplanar monopole antenna for passive chipless UWB-RFID localization system. *IEEE Transactions on Antennas and Propagation*, 58, 271–278.

Huang, S. and Wu, J. (2002). Optical watermark. US20020054680A1.

Huang, S. and Wu, J.K. (2007). Optical watermarking for printed document authentication. *IEEE Transactions on Information Forensics and Security*, 2, 164–173.

Hutchings, I.M. and Martin, G.D. (eds) (2012). *Inkjet Technology for Digital Fabrication*. John Wiley & Sons, New York.

Hutton, R.G. and Merry, T. (1977). Documents of value including intaglio printed transitory images. US4033059A.

Islam, M.A., Yap, Y., Karmakar, N., and Azad, A.K.M. (2012). Orientation independent compact chipless RFID tag. In *Proceedings of the IEEE International Conference on RFID-Technologies and Applications (RFID-TA)*, 137–141, https://doi.org/10.1109/RFID-TA.2012.6404499.

Jayakrishnan, M.P., Perret, E., and Vena, A. (2020). *Non-Volatile CBRAM/MIM Switching Technology for Electronically Reconfigurable Passive Microwave Devices*. ISTE Press, London and Elsevier, Oxford.

Jeena, C.S. (2014). Steps to identify authentication solutions to curb counterfeiting. *The Holography Times*, 7, 6–9.

Jung, H.C., Cho, S.-H., Joung, J.W., and Oh, Y.-S. (2007). Studies on inkjet-printed conducting lines for electronic devices. *Journal of Electronic Materials*, 36, 1211–1218.

Kamaludin, H., Mahdin, H., and Abawajy, J.H. (2018). Clone tag detection in distributed RFID systems. *PLoS One*, 13, e0193951, https://doi.org/10.1371/journal.pone.0193951.

Kerns, W.J., Brogger, B., and Esterberg, J.L. (2006). Identification particles and system and method for retrospective identification using spectral codes. US7038766B2.

Khan, M.M., Tahir, F.A., Farooqui, M.F., Shamim, A., and Cheema, H.M. (2016). 3.56-bits/cm^2 compact inkjet printed and application specific chipless RFID tag. *IEEE Antennas and Wireless Propagation Letters*, 15, 1109–1112.

Khandpur, R.S. (2006). *Printed Circuit Boards: Design, Fabrication, Assembly and Testing*. McGraw-Hill Companies, New York.

Kheir, M., Kreft, H., Hölken, I., and Knöchel, R. (2014). On the physical robustness of RF on-chip nanostructured security. *Journal of Information Security and Applications*, 19, 301–307, https://doi.org/10.1016/j.jisa.2014.09.007.

Koltai, F., Baros, L., Adam, B., and Takacs, F. (2000). Anti-counterfeiting method and apparatus using digital screening. US6104812A.

Kracek, J., Svanda, M., and Hoffmann, K. (2019). Scalar method for reading of chipless RFID tags based on limited ground plane backed dipole resonator array. *IEEE Transactions on Microwave Theory and Techniques*, 67, 4547–4558.

Krustok, J., Collan, H., Yakushev, M., and Hjelt, K. (1999). The role of spatial potential fluctuations in the shape of the PL bands of multinary semiconductor compounds. *Physica Scripta*, T79, 179.

Lakafosis, V., Traille, A., Lee, H., Gebara, E., Tentzeris, M.M., Dejean, G.R., and Kirovski, D. (2011). RF fingerprinting physical objects for anticounterfeiting applications. *IEEE Transactions on Microwave Theory and Techniques*, 59, 504–514.

Lawandy, N.M. (1998). Field activated security thread including polymer dispersed liquid crystal. WO1998036376A1.

Lawandy, N. and Tillotson, S. (2001). Authentication using a digital watermark. US20010037455A1.

Lawandy, N., Smuk, A., Olson, L., and Zepp, C. (2014). Multi wavelength excitation/emission authentication and detection scheme. EP2805288A1.

Lázaro, A., Villarino, R., Costa, F., Genovesi, S., Gentile, A., Buoncristiani, L., and Girbau, D. (2018). Chipless dielectric constant sensor for structural health testing. *IEEE Sensors Journal*, 18, 5576–5585, https://doi.org/10.1109/JSEN.2018.2839689.

Lehtonen, M., Staake, T., and Michahelles, F. (2008). *Networked RFID Systems and Lightweight Cryptography: Raising Barriers to Product Counterfeiting*. Springer, Berlin, Heidelberg, 169–187.

Levy, K., Labrec, B., and Jones, R. (2005). Color laser engraving and digital watermarking. US20050001419A1.

Li, L. (2013). Technology designed to combat fakes in the global supply chain. *Business Horizons, Special Issue: Protecting your Intellectual Property Rights*, 56, 167–177.

Loving, C.D. (1995). Microtag and method. US5411799A.

Lu, F., Wang, H., Guo, Y., Tan, Q., Zhang, W., and Xiong, J. (2018). Microwave backscatter-based wireless temperature sensor fabricated by an alumina-backed Au slot radiation patch. *Sensors*, 18, 242, https://doi.org/10.3390/s18010242.

Mallik, D.W. (1992). Holograms with discontinuous metallization including alpha-numeric shapes. US5142383A.

Maltoni, D., Maio, D., Jain, A., and Prabhakar, S. (2009). *Handbook of Fingerprint Recognition*, 2nd edition. Springer-Verlag, London.

Marchand, J.F.P. (2003). Microwave barcode reader using dipole antenna. US6547140B2.

Marindra, A.M.J. and Tian, G.Y. (2018). Chipless RFID sensor tag for metal crack detection and characterization. *IEEE Transactions on Microwave Theory and Techniques*, 66, 2452–2462, https://doi.org/10.1109/TMTT.2017.2786696.

Matbouly, H.E., Tedjini, S., Zannas, K., and Duroc, Y. (2018). Compact multi bit slotted C-scatterer for threshold sensitive chipless wireless temperature sensor. *Technologies*, 6, 59.

Mazaika, P.K. (2004). Method for invisible embedded data using yellow glyphs. US6708894B2.

McCarthy, L. and Swiegers, G. (2007). Method of encoding a latent image. US20070121170A1.

McGrew, S.P. (1995). Anticounterfeiting method and device utilizing holograms and pseudorandom dot patterns. US5396559A.

Mercolino, T.J. (2013). Nucleic acid-based authentication codes. WO2013138300A1.

Merilampi, S.L., Björninen, T., Vuorimaki, A., Ukkonen, L., Ruuskanen, P., and Sydänheimo, L. (2010). The effect of conductive ink layer thickness on the functioning of printed UHF RFID antennas. *Proceedings of the IEEE*, 98, 1610–1619.

Moore, L.J. (1997). Anti-counterfeiting system. US5592561A.

Nair, R., Barahona, M.,Betancourt, D., Cornelius Schmidt, G., Bellmann, M., Hoft, D., Plettemeier, D., Hubler, A., and Ellinger, F. (2014a). A fully printed passive chipless RFID tag for low-cost mass production. In *Proceedings of the 8th European Conference on Antennas and Propagation (EuCAP 2014)*, 2950–2954.

Nair, R., Barahona, M., Betancourt, D., Schmidt, G., Bellmann, M., Hoft, D., Plettermeier, D., Hubler, A., and Ellinger, F. (2014b). A novel fully printed 28-bits capacity chipless RFID tag based on open conical resonators. In *Proceedings of the Progress in Electromagnetics Research Symposium*, Guangzhou, China, p. 5.

Nemeth, J. (2002). Security document with raised intaglio printed image. EP1194302A1.

Nur, H.M., Song, J.H., Evans, J.R.G., and Edirisinghe, M.J. (2002). Ink-jet printing of gold conductive tracks. *Journal of Materials Science: Materials in Electronics* 13, 213–219.

Obama, B. (2011). *Strategy to Combat Transnational Organized Crime*. White House.

OECD. (2009). *Magnitude of Counterfeiting and Piracy of Tangible Products: An Update*. Organisation for Economic Co-operation and Development, Paris.

OECD, EUIPO. (2016). *Trade in Counterfeit and Pirated Goods*. Organisation for Economic Co-operation and Development, Paris.

Outwater, C. and Helmick, M. (2002). Composite authentication mark and system and method for reading the same. US6354501B1.

Outwater, C. and Loop, R. (2001). Label having an invisible bar code applied thereon. US6203069B1.

Paggi, C., Occhiuzzi, C., and Marrocco, G. (2014). Sub-millimeter displacement sensing by passive UHF RFID antennas. *IEEE Transactions on Antennas and Propagation*, 62, 905–912.

Pecht, M. and Ganesan, S. (2006). Open trace defects in FR4 printed circuit boards. *Circuit World*, 32, 3–7.

Periaswamy, S.C.G., Thompson, D.R., and Di, J. (2011). Fingerprinting RFID tags. *IEEE Transactions on Dependable and Secure Computing*, 8, 938–943.

Perret, E. (2014). *Radio Frequency Identification and Sensors: From RFID to Chipless RFID*. John Wiley & Sons, New York.

Perret, E. (2016). Permittivity characterization based on Radar Cross measurements. In *Proceedings of the URSI International Symposium on Electromagnetic Theory (EMTS)*, 457–460.

Perret, E. (2017). Displacement sensor based on radar cross-polarization measurements. *IEEE Transactions on Microwave Theory and Techniques*, 65, 955–966.

Perret, E., Hamdi, M., Tourtollet, G.E.P., Nair, R., Garet, F., Delattre, A., Vena, A., Duvillaret, L., Martinez, P., Tedjini, S., and Boutant, Y. (2013). THID, the next step of chipless RFID. *Proceedings of the IEEE International Conference on RFID*, Florida, USA, 261–268.

Phillips, G.K. (2004). Security document with nano-pattern. US6692030B1.

Phillips, G. and Phillips, N. (2005). Copy-resistant security paper. US20050086 482A1.

Pizzanelli, D.J. (1997). Holograms for security markings. US5623347A.

Power, G. (2008). Anti-counterfeit Technologies for the Protection of Medicines. World Health Organization, Geneva, Switzerland.

Pynttari, V., Halonen, E., Sillanpaa, H., Mantysalo, M., and Mantysalo, M. (2012). RF design for inkjet technology: antenna geometries and layer thickness optimization. *IEEE Transactions on Antennas and Propagation*, 11, 188–191.

Raksha, V.P., Delst, C.J., Coombs, P.G., Markantes, C.T., and Argoitia, A., (2015). Printed magnetic ink overt security image. USRE45762E1.

Ramos, A., Perret, E., Rance, O., Tedjini, S., Lazaro, A., and Girbau, D. (2016). Temporal separation detection for chipless depolarizing frequency-coded RFID. *IEEE Transactions on Microwave Theory and Techniques*, 64, 2326–2337.

Rance, O., Siragusa, R., Lemaitre-Auger, P., and Perret, E. (2016). Toward RCS magnitude level coding for chipless RFID. *IEEE Transactions on Microwave Theory and Techniques*, 64, 2315–2325.

Rance, O., Perret, E., Siragusa, R., and Lemaitre-Auger, P. (2017). *RCS Synthesis for Chipless RFID: Theory and Design*. ISTE Press, London and Elsevier, Oxford.

Rao, K.V.S., Nikitin, P.V., and Lam, S.F. (2005). Antenna design for UHF RFID tags: A review and a practical application. *IEEE Transactions on Antennas and Propagation*, 53, 3870–3876, https://doi.org/10.1109/TAP.2005.859919.

Redinger, D., Molesa, S., Yin, S., Farschi, R., and Subramanian, V. (2004). An ink-jet-deposited passive component process for RFID. *IEEE Transactions Electron Devices*, 51, 1978–1983.

Rennane, A., Abdelnour, A., Kaddour, D., Touhami, R., and Tedjini, S. (2018). Design of passive UHF RFID sensor on flexible foil for sports balls pressure monitoring. *IET Microwaves, Antennas & Propagation*, 12(14), 2154–2160.

Rezaiesarlak, R. and Manteghi, M. (2013). Short-time matrix pencil method for chipless RFID detection applications. *IEEE Transactions on Antennas and Propagation*, 61, 2801–2806.

Rezaiesarlak, R. and Manteghi, M. (2014a). Complex-natural-resonance-based design of chipless RFID tag for high-density data. *IEEE Transactions on Antennas and Propagation*, 62, 898–904.

Rezaiesarlak, R. and Manteghi, M. (2014b). A space-frequency technique for chipless RFID tag localization. *IEEE Transactions on Antennas and Propagation*, 62, 5790–5797.

Rezaiesarlak, R. and Manteghi, M. (2015a). Accurate extraction of early-/late-time responses using short-time matrix pencil method for transient analysis of scatterers. *IEEE Transactions on Antennas and Propagation*, 63, 4995–5002.

Rezaiesarlak, R. and Manteghi, M. (2015b). Design of chipless RFID tags based on characteristic mode theory (CMT). *IEEE Transactions on Antennas and Propagation*, 63, 711–718.

RFIDTECH large reader [Online]. Available: http://www.rfidtechgroup.com/home/product.html (Accessed 18 January 2019).

Rhoads, G.B. (2013). Authentication using a digital watermark. US8505108B2.

Romero, H.P., Remley, K.A., Williams, D.F., and Wang, C.M. (2009). Electromagnetic measurements for counterfeit detection of radio frequency identification cards. *IEEE Transactions on Microwave Theory and Techniques*, 57, 1383–1387.

Romero, H.P., Remley, K.A., Williams, D.F., Wang, C.-M., and Brown, T.X. (2010). Identifying RF identification cards from measurements of resonance and carrier harmonics. *IEEE Transactions on Microwave Theory and Techniques*, 58, 1758–1765.

Romero, F.J.J., García, J.V., Fernández, L.J.F., Gamo, A.J., and Rodríguez, B.M.Á. (2017). Use of radiofrequency wave absorbing markers for the authentication of security documents. EP2694600B1.

Royer, A. (2001). Printed document including bar code authentication system. US6308991B1.

Sajitha, V.R., Nijas, C. M., Roshna, T. K., Vasudevan, K., and Mohanan, P. (2016). Compact cross loop resonator based chipless RFID tag with polarization insensitivity. *Microwave and Optical Technology Letters*, 58, 944–947.

Salmerón, J.F., Molina-Lopez, F., Briand, D., Ruan, J.J., Rivadeneyra, A., Carvajal, M.A., Capitán-Vallvey, L.F., de Rooij, N.F., and Palma, A.J. (2014a). Properties and printability of inkjet and screen-printed silver patterns for RFID antennas. *Journal of Electronic Materials*, 43, 604–617.

Salmerón, J.F., Molina-Lopez, F., Rivadeneyra, A., Quintero, A.V., Capitán-Vallvey, L.F., de Rooij, N.F., Ozáez, J.B., Briand, D., and Palma, A.J. (2014b). Design and development of sensing RFID tags on flexible foil compatible with EPC Gen 2. *IEEE Sensors Journal*, 14, 4361–4371.

Sarkar, T.K. and Pereira, O. (1995). Using the matrix pencil method to estimate the parameters of a sum of complex exponentials. *IEEE Antennas and Propagation Magazine*, 37, 48–55.

Sarrazin, F., Chauveau, J., Pouliguen, P., Potier, P., and Sharaiha, A. (2014). Accuracy of Singularity expansion method in time and frequency domains to characterize antennas in presence of noise. *IEEE Transactions on Antennas and Propagation*, 62, 1261–1269.

Seifert, H.A. (1999). Security paper having an embedded security thread and a process for making the security paper. US5944927A.

Shao, B., Chen, Q., Amin, Y., Hllstedt, J., Liu, R., Tenhunen, H., and Zheng, L.-R. (2009). Process-dependence of inkjet printed folded dipole antenna for 2.45 GHz RFID tags. In *Proceedings of the 3rd European Conference on Antennas and Propagation*, 2336–2339.

Sheu, J.-J., Chen, L.-L., Chen, E., and Liang, B. (2002). Method of utilizing ribonucleic acid as markers for product anti-counterfeit labeling and verification. US20020187263A1.

Sheu, J.-J., Chen, L.-L., Chen, E., and Liang, B. (2006). Method of marking solid or liquid substances with nucleic acid for anti-counterfeiting and authentication. US7115301B2.

Smith, K.J., Vahey, P.G., Fraenkel, H.A., Bray, R.G., Lytle, N.W., Rapoport, W.R., and Smith, G.A. (2012). Authenticatable mark, systems for preparing and authenticating the mark. US8330122B2.

Stierman, T., Hunt, D., and Zdon, C. (2008). Three-dimensional authentication of microparticle mark. US20080034426A1.

Tan, Y. (1997). Fluorescent security system for printed transaction records. US5667317A.

Toyoda, K., Mathiopoulos, P.T., Sasase, I., and Ohtsuki, T. (2017). A novel blockchain-based product ownership management system (POMS) for anti-counterfeits in the post supply chain. *IEEE Access*, 5, 17465–17477, https://doi.org/10.1109/ACCESS.2017.2720760.

Turki, B.M., Parker, E.A., Ali Ziai, M., Batchelor, J.C., Sanchez-Romaguera, V., and Yeates, S.G. (2013). Study of printing errors in digitally fabricated FSS. In *Proceedings of the Loughborough Antennas & Propagation Conference (LAPC)*, 429–432.

Vena, A., Perret, E., and Tedjini, S. (2011). Chipless RFID tag using hybrid coding technique. *IEEE Transactions on Microwave Theory and Techniques*, 59, 3356–3364.

Vena, A., Perret, E., and Tedjini, S. (2012a). Design of compact and auto-compensated single-layer chipless RFID tag. *IEEE Transactions on Microwave Theory and Techniques*, 60, 2913–2924.

Vena, A., Perret, E., and Tedjini, S. (2012b). A fully printable chipless RFID tag with detuning correction technique. *IEEE Microwave and Wireless Components Letters*, 22, 209–211.

Vena, A., Perret, E., and Tedjini, S. (2012c). High-Capacity chipless RFID tag insensitive to the polarization. *IEEE Transactions on Antennas and Propagation*, 60, 4509–4515.

Vena, A., Babar, A.A., Sydanheimo, L., Tentzeris, M.M., and Ukkonen, L. (2013a). A novel near-transparent ASK-reconfigurable inkjet-printed chipless RFID Tag. *IEEE Transactions on Antennas and Propagation*, 12, 753–756.

Vena, A., Perret, E., Tedjini, S., Eymin Petot Tourtollet, G., Delattre, A., Garet, F., and Boutant, Y. (2013b). Design of chipless RFID tags printed on paper by flexography. *IEEE Transactions on Antennas and Propagation*, 61, 5868–5877.

Vena, A., Perret, E., and Tedjni, S. (2013c). A depolarizing chipless RFID tag for robust detection and its FCC compliant UWB reading system. *IEEE Transactions on Microwave Theory and Techniques*, 61, 2982–2994.

Vena, A., Sorli, B., Foucaran, A., and Belaizi, Y. (2014a). A RFID-enabled sensor platform for pervasive monitoring. In *Proceedings of the 9th International Symposium on Reconfigurable and Communication-Centric Systems-on-Chip (ReCoSoC)*, 1–4.

Vena, A., Tedjini, M., Björninen, T., Sydänheimo, L., Ukkonen, L., and Tentzeris, M.M. (2014b). A novel inkjet-printed wireless chipless strain and crack sensor on flexible laminates. In *Proceedings of the IEEE Antennas and Propagation Society International Symposium (APSURSI)*, 1294–1295.

Vena, A., Perret, E., Sorli, B., and Tedjini, S. (2015a). Theoretical study on detection distance for chipless RFID systems according to transmit power regulation standards. In *Proceedings of the 9th European Conference on Antennas Propagation (EuCAP)*, 1–4.

Vena, A., Sydänheimo, L., Tentzeris, M.M., and Ukkonen, L. (2015b). A fully inkjet-printed wireless and chipless sensor for CO_2 and temperature detection. *IEEE Sensors Journal*, 15, 89–99.

Vena, A., Perret, E., Kaddour, D., and Baron, T. (2016a). Toward a reliable chipless RFID humidity sensor tag based on silicon nanowires. *IEEE Transactions on Microwave Theory and Techniques*, 64, 2977–2985.

Vena, A., Perret, E., and Tedjini, S. (2016b). *Chipless RFID based on RF Encoding Particle: Realization, Coding and Reading System*. ISTE Press, London and Elsevier, Oxford.

Vig, R. and Saglimbeni, A. (2005). Tamper-resistant authentication mark for use in product or product packaging authentication. US20050243305A1.

Welle, R.P. (1998). Isotopic taggant method and composition. US5760394A.

Wild, H. and Brehm, L. (2005). Optically variable element comprising a sequence of thin-film layers. US20050141094A1.

Yang, C. and Sample, A.P. (2016). EM-ID: Tag-less identification of electrical devices via electromagnetic emissions. In *Proceedings of the IEEE International Conference on RFID*, Florida, MA, USA, 1–8.

Yang, L., Peng, P., Dang, F., Wang, C., Li, X.-Y., and Liu, Y. (2015). Anti-counterfeiting via federated RFID tags' fingerprints and geometric relationships. In *Proceedings of the IEEE Conference on Computer Communications (INFOCOM)*, Kowloon, Hong Kong, 1966–1974.

Yang, K., Forte, D., and Tehranipoor, M.M. (2016). UCR: An unclonable chipless RFID tag. In *Proceedings of the IEEE International Symposium on Hardware Oriented Security Trust (HOST)*, Washington, DC, USA, 7–12.

Zanetti, D., Fellmann, L., Capkun, S., and Čapkun, S. (2010). Privacy-preserving clone detection for RFID-enabled supply chains. In *Proceedings of the IEEE International Conference on RFID (IEEE RFID 2010)*, 37–44.

Zannas, K., Matbouly, H.E., Duroc, Y., and Tedjini, S. (2018). Self-tuning RFID tag: A new approach for temperature sensing. *IEEE Transactions on Microwave Theory and Techniques*, 66, 5885–5893.

Zhang, G., Xia, L., Jia, S., and Ji, Y. (2016). Identification of Cloned HF RFID Proximity Cards Based on RF Fingerprinting. In *Proceedings of the IEEE Trustcom/BigDataSE/ISPA*, Tianjin, China, 292–300.

Index

Other titles from

iSTE

in

Electronics Engineering

2022

METHAPETTYPARAMBU PURUSHOTHAMA Jayakrishnan, PERRET Etienne, VENA Arnaud
Non-Volatile CBRAM/MIM Switching Technology for Electronically Reconfigurable Passive Microwave Devices: Theory and Methods for Application in Rewritable Chipless RFID

2021

NGUYEN Thien-Phap
Organic Electronics 1: Materials and Physical Processes
Organic Electronics 2: Applications and Marketing

2020

CAPPY Alain
Neuro-inspired Information Processing

FOUQUET François
Noise in Radio-Frequency Electronics and its Measurement

MAHDOUM Ali
CAD of Circuits and Integrated Systems

2018

MURET Pierre
Fundamentals of Electronics 2: Continuous-time Signals and Systems
Fundamentals of Electronics 3: Discrete-time Signals and Systems, and Quantized Level Systems

2017

MURET Pierre
Fundamentals of Electronics 1: Electronic Components and Elementary Functions

BUCCI Davide
Analog Electronics for Measuring Systems

2016

BAUDRAND Henri, TITAOUINE Mohammed, RAVEU Nathalie
The Wave Concept in Electromagnetism and Circuits: Theory and Applications

FANET Hervé
Ultra Low Power Electronics and Adiabatic Solutions

NDJOUNTCHE Tertulien
Digital Electronics 1: Combinational Logic Circuits
Digital Electronics 2: Sequential and Arithmetic Logic Circuits
Digital Electronics 3: Finite-state Machines

2015

DURAFFOURG Laurent, ARCAMONE Julien
Nanoelectromechanical Systems

2014

APPRIOU Alain
Uncertainty Theories and Multisensor Data Fusion

CONSONNI Vincent, FEUILLET Guy
Wide Band Gap Semiconductor Nanowires 1: Low-Dimensionality Effects and Growth
Wide Band Gap Semiconductor Nanowires 2: Heterostructures and Optoelectronic Devices

GAUTIER Jean-Luc
Design of Microwave Active Devices

LACAZE Pierre Camille, LACROIX Jean-Christophe
Non-volatile Memories

TEMPLIER François
OLED Microdisplays: Technology and Applications

THOMAS Jean-Hugh, YAAKOUBI Nourdin
New Sensors and Processing Chain

2013

COSTA François, GAUTIER Cyrille, LABOURE Eric, REVOL Bertrand
Electromagnetic Compatibility in Power Electronics

KORDON Fabrice, HUGUES Jérôme, CANALS Agusti, DOHET Alain
Embedded Systems: Analysis and Modeling with SysML, UML and AADL

LE TIEC Yannick
Chemistry in Microelectronics

2012

BECHERRAWY Tamer
Electromagnetism: Maxwell Equations, Wave Propagation and Emission

LALAUZE René
Chemical Sensors and Biosensors

LE MENN Marc
Instrumentation and Metrology in Oceanography

SAGUET Pierre
Numerical Analysis in Electromagnetics: The TLM Method

2011

ALGANI Catherine, RUMELHARD Christian, BILLABERT Anne-Laure
Microwaves Photonic Links: Components and Circuits

BAUDRANT Annie
Silicon Technologies: Ion Implantation and Thermal Treatment

DEFAY Emmanuel
Integration of Ferroelectric and Piezoelectric Thin Films: Concepts and Applications for Microsystems

DEFAY Emmanuel
Ferroelectric Dielectrics Integrated on Silicon

BESNIER Philippe, DÉMOULIN Bernard
Electromagnetic Reverberation Chambers

LANDIS Stefan
Nano-lithography

2010

LANDIS Stefan
Lithography

PIETTE Bernard
VHF / UHF Filters and Multicouplers

2009

DE SALVO Barbara
Silicon Non-volatile Memories / Paths of Innovation

DECOSTER Didier, HARARI Joseph
Optoelectronic Sensors

FABRY Pierre, FOULETIER Jacques
Chemical and Biological Microsensors / Applications in Fluid Media

GAUTIER Jacques
Physics and Operation of Silicon Devices in Integrated Circuits

MOLITON André
Solid-State Physics for Electronics

PERRET Robert
Power Electronics Semiconductor Devices

SAGUET Pierre
Passive RF Integrated Circuits

2008

CHARRUAU Stéphane
Electromagnetism and Interconnections

2007

RIPKA Pavel, TIPEK Alois
Modern Sensors Handbook

Printed and bound by CPI Group (UK) Ltd, Croydon, CR0 4YY
13/09/2026
14961042-0001